Capturing the Multiple Benefits of Energy Efficiency

INTERNATIONAL ENERGY AGENCY

The International Energy Agency (IEA), an autonomous agency, was established in November 1974. Its primary mandate was – and is – two-fold: to promote energy security amongst its member countries through collective response to physical disruptions in oil supply, and provide authoritative research and analysis on ways to ensure reliable, affordable and clean energy for its 29 member countries and beyond. The IEA carries out a comprehensive programme of energy co-operation among its member countries, each of which is obliged to hold oil stocks equivalent to 90 days of its net imports. The Agency's aims include the following objectives:

- Secure member countries' access to reliable and ample supplies of all forms of energy; in particular, through maintaining effective emergency response capabilities in case of oil supply disruptions.
- Promote sustainable energy policies that spur economic growth and environmental protection in a global context – particularly in terms of reducing greenhouse-gas emissions that contribute to climate change.
- Improve transparency of international markets through collection and analysis of energy data.
- Support global collaboration on energy technology to secure future energy supplies and mitigate their environmental impact, including through improved energy efficiency and development and deployment of low-carbon technologies.
- Find solutions to global energy challenges through engagement and dialogue with non-member countries, industry, international organisations and other stakeholders.

IEA member countries:

Australia
Austria
Belgium
Canada
Czech Republic
Denmark
Estonia
Finland
France
Germany
Greece
Hungary
Ireland
Italy
Japan
Korea (Republic of)
Luxembourg
Netherlands
New Zealand
Norway
Poland
Portugal
Slovak Republic
Spain
Sweden
Switzerland
Turkey
United Kingdom
United States

The European Commission also participates in the work of the IEA.

40 iea International Energy Agency 1974•2014

Secure • Sustainable • Together

© OECD/IEA, 2014

International Energy Agency
9 rue de la Fédération
75739 Paris Cedex 15, France

www.iea.org

Please note that this publication is subject to specific restrictions that limit its use and distribution.
The terms and conditions are available online at
http://www.iea.org/termsandconditionsuseandcopyright/

Table of contents

Introduction 7

Foreword 7
Acknowledgements 8
Executive summary 18

Chapter 1 Taking a multiple benefits approach to energy efficiency 27

Introducing the multiple benefits of energy efficiency 27
Energy efficiency: From "hidden fuel" to "first fuel" 29
A multiple benefits approach adds value to energy efficiency 33
The rebound effect in a multiple benefits context 38
Empowering policy makers to assess multiple benefits 39
Conclusion 41
Bibliography 41

Chapter 2 Macroeconomic impacts of energy efficiency 45

Introduction: Increasing evidence of macroeconomic benefits 45
Framing the macroeconomic impacts 46
Sample results for key macroeconomic impacts 52
Methodological approaches 56
Policy-making considerations 62
Further research for stakeholders 65
Conclusions 66
Bibliography 66

Chapter 3 Public budget impacts of energy efficiency 71

Introduction: Emerging evidence of public budget benefits 71
The range of public budget impacts 73
Methodological approaches 84
Policy-making considerations 90
Further research for stakeholders 91
Conclusions 92
Bibliography 93

Chapter 4 Health and well-being impacts of energy efficiency 97

Introduction: Linking energy efficiency and health 97
The range of health and well-being impacts 100

© OECD/IEA, 2014.

Methodological approaches 109
Policy-making considerations 116
Further research for stakeholders 118
Conclusions 118
Bibliography 119

Chapter 5 Industrial sector impacts of energy efficiency 129

Introduction: Emerging evidence of industrial sector impacts 129
The range of industrial sector impacts 134
Methodological approaches 137
Policy-making considerations 146
Further research for stakeholders 148
Conclusions 149
Bibliography 149

Chapter 6 Energy delivery impacts of energy efficiency 153

Introduction: A strong evidence base for energy delivery impacts 153
The range of energy delivery impacts 156
Methodological approaches 170
Further research for stakeholders 174
Conclusions 174
Bibliography 175

Chapter 7 Conclusion: A new perspective on energy efficiency 179

Introduction 179
Optimising the multiple benefits approach 180
Interdisciplinary co-operation 181
Conclusion 186
Bibliography 187

Companion Guide Methodologies for the multiple benefits approach 189

The existing context for energy efficiency policy assessment 190
Measuring the multiple benefits 191
Using multiple benefits values for policy decision making 202
Conclusions 208
Bibliography 208

Annexes 212

Annex A Glossary 213
Annex B Acronyms, abbreviations and units 216
Annex C List of boxes, figures and tables 221

© OECD/IEA, 2014.

List of rebound effect perspectives

Chapter 1 Taking a multiple benefits approach to energy efficiency 27

Rebound effect perspective 1: Understanding the dynamics of the rebound effect: When does it matter? 40

Chapter 2 Macroeconomic impacts of energy efficiency 45

Rebound effect perspective 2: The macroeconomic rebound effect 64

Chapter 3 Public budget impacts of energy efficiency 71

Rebound effect perspective 3: Analysis of public budget impacts 90

Chapter 4 Health and well-being impacts of energy efficiency 97

Rebound effect perspective 4: Analysis of health and well-being impacts 106

Chapter 5 Industrial sector impacts of energy efficiency 129

Rebound effect perspective 5: Analysis of industrial sector impacts 141

Chapter 6 Energy delivery impacts of energy efficiency 153

Rebound effect perspective 6: Analysis in the context of energy utilities 173

© OECD/IEA, 2014.

© OECD/IEA, 2014.

Foreword

I am pleased to release this report on the *Multiple Benefits of Energy Efficiency*. While energy security is foremost in our minds, and energy efficiency contributes to energy security, it can also give us the opportunity to grow our economies, provide jobs and increase the number of people with access to energy as well as a range of other benefits. This publication is our first attempt to bring together methodologies for assessing the impacts of a range of benefits that energy efficiency brings. Policy makers around the world have their own priorities for economic growth and social development and energy efficiency can contribute to many of them.

I see this report as being the first, essential step on the journey to assist policy makers in evaluating the impacts of both proposed and implemented energy efficiency measures in a range of ways that suit their own unique circumstances and objectives. This in turn will provide feedback to refine ongoing programmes and help design and prioritise future energy efficiency actions. It also helps to address the challenge of the lack of visibility of energy efficiency by acknowledging and valuing the positive impacts it triggers across a variety of areas.

While it can be difficult to monetise, quantify or even measure these broader benefits, this publication highlights early achievements in doing precisely that. In some cases, including the value of multiple benefits alongside traditional benefits has shown energy efficiency measures delivering returns as high as EUR 4 for every EUR 1 invested. Case studies throughout the publication confirm a solid foundation from which to launch further expansion of the multiple benefits approach to energy efficiency. What is more, the publication demonstrates that the rebound effect, which sees people or businesses redirecting energy savings towards other energy-consuming activities, is not necessarily a negative outcome – it can be a good thing from the perspective of contributing to economic and social objectives.

This publication is the result of an extensive review and synthesis of the state of the art of quantifying the multiple benefits of energy efficiency. More than 300 people from 27 countries and over 60 organisations were involved in the process and we believe this publication will lead to increased efforts to refine metrics and tools, build confidence in the multiple benefits approach and encourage innovation in energy efficiency policy design.

This report is published under my authority as Executive Director of the IEA.

Maria van der Hoeven,
Executive Director
International Energy Agency

© OECD/IEA, 2014.

Acknowledgements

This study was prepared by the Energy Efficiency Unit, Directorate for Sustainable Policy and Technology, of the International Energy Agency (IEA).

The project leader and main author of this study is Nina Campbell. The IEA Multiple Benefits project was initiated by Lisa Ryan, previously of the IEA, who was also the lead author on Chapters 2 and 3, and provided general inputs and guidance to the study. Additional contributing authors were Vida Rozite, Eoin Lees (Regulatory Assistance Project) and Grayson Heffner (previously of the IEA).

The author extends special thanks to key experts who played advisory roles in drafting the focus benefit chapters:

Chapter 2. Macroeconomic development and **Chapter 3. Public budgets**: Sigurd Naess-Schmidt (Copenhagen Economics) and Hector Pollitt (Cambridge Econometrics).

Chapter 4. Health and well-being: Philippa Howden-Chapman (University of Otago, New Zealand), Christine Liddell (University of Ulster, United Kingdom) and Jim Scheer (Sustainable Energy Authority of Ireland).

Chapter 5: Industrial productivity: Amy Philbrook (Department of Industry, government of Australia).

Chapter 6. Energy delivery: John Appleby (Natural Resources Canada, government of Canada), Jennifer Kallay (Synapse Energy Economics, Inc.), Matthew Lam (Natural Resources Canada, government of Canada), Joy Morgenstern (California Public Utilities Commission, United States) and Tim Woolf (Synapse Energy Economics, Inc.).

For their leadership and guidance, the author would like to thank Maria van der Hoeven, Executive Director of the IEA, Didier Houssin, Director of the Sustainable Policy and Technology Directorate, Philippe Benoit, Head of the Energy Efficiency and Environment Division, Robert Tromop, Head of the Energy Efficiency Unit, and Rebecca Gaghen, Head of the Communication and Information Office. Thanks are extended to various IEA colleagues who contributed inputs and feedback on drafts, including: Sara Bryan-Pasquier, Tyler Bryant, Charlotte Forbes, Timur Gul, Kyeong Hee Je, Lorcan Lyons, Araceli Pales Fernandez, Uwe Remme, Melanie Slade, Kira West and Matthew Wittenstein. Special thanks to Muriel Custodio, Astrid Dumond, Therese Walsh and Bertrand Sadin from the Communication and Information Office for support in the publication process, Rachael Boyd and Elizabeth Spong from the Office of Legal Counsel, and Greg Frost for press support.

Marilyn Smith was the managing editor of this publication, with support from Jessica Hutchings.

Workshops

A series of Expert Roundtables and an extensive consultation and review process were held to gather input to this study, resulting in valuable new information, insights and feedback:

- Macroeconomic Impacts of Energy Efficiency Improvements: Paris, 24 January 2013
- Health & Well-being Impacts of Energy Efficiency Improvements: Copenhagen, 18 April 2013

© OECD/IEA, 2014.

- Energy Provider and Consumer Benefits of Energy Efficiency: Ottawa, 15 October 2013
- Industrial Productivity and Competitiveness Benefits of Energy Efficiency: Paris, 27 January 2014

Workshop presentations and more information are available at: www.iea.org/topics/energyefficiency/multiplebenefitsofenergyefficiency/

The Multiple Benefits Network

Over 300 international experts participated in this research and consultation process. Those people now make up part of the Multiple Benefits Network established by the IEA in the context of this study. The Multiple Benefits Network will continue to facilitate information sharing with regards to policy and research on multiple benefits and support the building of consensus on assessment methods.

To join this network, please contact: nina.campbell@iea.org

Macroeconomic development

John Appleby	Natural Resources Canada, government of Canada
Bogdan Atanasiu	Buildings Performance Institute Europe (BPIE)
Mourad Ayouz	*Électricité de France* (EDF)
Robert Ayres	INSEAD, France
Peter Bach	Danish Energy Agency, government of Denmark
Terry Barker	University of Cambridge, United Kingdom
François Cohen	London School of Economics (LSE), United Kingdom
Anca-Diana Barbu	European Environment Agency (EEA)
Ron Benioff	National Renewable Energy Laboratory (NREL), United States
Kornelis Blok	Ecofys, the Netherlands
Martin Bo Hansen	Copenhagen Economics, Denmark
Bjørg Bogstrand	Permanent Delegation to the OECD, government of Norway
Randall Bowie	Rockwool International A/S, United States
Ioulia Boussel	APBE Energy Forecasting Agency, Russia
Gaël Callonnec	*Agence de l'environnement et de la maîtrise de l'énergie* (*Ademe*) (Agency for Environment and Energy Management), government of France
Unnada Chewpreecha	Cambridge Econometrics, United Kingdom
Laurence Cheyrou	*Ministère l'écologie, du développement durable, des transports et du logement* (*MEDDTL*) (Ministry of Ecology, Sustainable Development, Transport and Housing), government of France
Jean Chateau	Environment Directorate, Organisation for Economic Co-operation and Development (OECD)
Rob Dellink	Environment Directorate, OECD
Benoit Dome	European Copper Alliance
Kristen Dyhr-Mikkelsen	Ea Energy Analyses, Denmark
Patty Fong	European Climate Foundation
Vivien Foster	World Bank

© OECD/IEA, 2014.

Roger Fouquet	LSE, United Kingdom
Frédéric Ghersi	*Centre international de recherche sur l'environnement et le Développement* (*CIRED*) (International Centre for Environment and Development Research), France
Peter Graham	Global Buildings Performance Network (GBPN), France
Nathalie Girouard	Green Growth and Sustainable Development Unit, OECD
Erik Gudbjerg	LokalEnergi, Denmark
Pedro Guertler	Association for the Conservation of Energy (ACE), United Kingdom
Christian Hederer	Permanent Delegation to the OECD, government of Austria
Khalil Heloui	EDF
Paul Hodson	Directorate-General for Energy, European Commission
Ingrid Holmes	E3G Third Generation Environmentalism Ltd, United Kingdom
Sven Iplikci	Cambridge Centre for Climate Change Research, United Kingdom
Rod Janssen	European Council for an Energy Efficient Economy (ECEEE)
Adrian Joyce	The European Alliance of Companies for Energy Efficiency in Buildings (EuroAce)
Yeonsoo Kim	Korea Development Institute, government of Korea
Youngil Kim	Korea Development Institute, government of Korea
Wilhelm Kuckshinrichs	Institute for Energy and Climate Research, Jülich Research Centre
Skip Laitner	American Council for Energy Efficient Economy (ACEEE)
Marie-Helene Laurent	EDF
Jens Lausten	GBPN
Ulrike Lehr	*Gesellschaft für wirtschaftliche Strukturforschung* (GWS mbH), Germany
Ines Lima Azevedo	Carnegie Mellon University, United States
Christian Lutz	GWS mbH, Germany
Leslie Malone	Environment Northeast, United States
Hannes MacNulty	Institute for Industrial Productivity (IIP)
Dorothy Maxwell	Global View Sustainability Services, Ltd.
Armin Mayer	Institute for Building Efficiency, United States
Niamh McDonald	GBPN
Edith Molenbroeck	Ecofys
Julian Morgan	European Central Bank
Annabelle Mourougane	Development Co-Operation Directorate, OECD
Sigurd Naess-Schmidt	Copenhagen Economics, Denmark
Karsten Neuhoff	German Institute for Economic Research (DIW Berlin)
Hector Pollitt	Cambridge Econometrics, United Kingdom
Prabodh Pourouchottamin	EDF
Frederic Reynes	*Observatoire français des conjonctures économiques* (*OFCE*) (French Economic Observatory), France

© OECD/IEA, 2014.

Oliver Rapf	BPIE
Harry Saunders	Decision Processes Incorporated, United States
Stefan Scheuer	Coalition for Energy Savings, Belgium
Laura Segafredo	Climate Works, United States
Mikael Skou Andersen	EEA
Lars Brømsøe Termanse	Danish Energy Authority, government of Denmark
Sam Thomas	Dept. of Energy & Climate Change, government of the United Kingdom
Johannes Thema	Wuppertal Institute for Climate, Environment and Energy (Wuppertal Institute), Germany
Stefan Thomas	Wuppertal Institute, Germany
Sergio Tirado-Hererro	Centre for Climate Change and Sustainable Energy Policy, Hungary
Karen Turner	Heriot-Watt University, Scotland
Prashant Vaze	Consultant

Public budgets

Dries Ake	European Climate
John Appleby	Natural Resources Canada, government of Canada
Bogdan Atanasiu	Buildings Performance Institute Europe (BPIE)
Martin Bo Hansen	Copenhagen Economics, Denmark
Jean Chateau	Environment Directorate, Organisation for Economic Co-operation and Development (OECD)
Unanda Chewpreecha	Cambridge Econometrics, United Kingdom
Kristen Dyhr-Mikkelsen	EA Energy Analyses, Denmark
Patty Fong	Climate Works, United States
Wilhelm Kuckshinrichs	*Forschungszentrum Juelich*, GmbH
Lorcan Lyons	International Energy Agency
Ulrike Lehr	*Gesellschaft für wirtschaftliche Strukturforschung* (GWS mbH), Germany
Christian Lutz	GWS mbH, Germany
Leslie Malone	Environment Northeast, United States
Hector Politt	Cambridge Econometrics, United Kingdom
Prabodh Pourouchottamin	*Électricité de France* (EDF)
Oliver Rapf	Building Performance Institute Europe
Harry Saunders	Decision Processes Incorporated, United States
Jim Scheer	Sustainable Energy Authority of Ireland (SEAI)
Mikael Skou Andersen	European Environment Agency (EEA)
Sam Thomas	Department of Energy & Climate Change, government of the United Kingdom
Stefan Thomas	Wuppertal Institute for Climate, Environment and Energy (Wuppertal Institute), Germany

© OECD/IEA, 2014.

Health and well-being

Roman Babut	National Energy Conservation Agency, government of Poland
Peter Bach	Danish Energy Authority, government of Denmark
Anca-Diana Barbu	European Environment Agency (EEA)
Martin Bo Hansen	Copenhagen Economics, Denmark
Brenda Boardman	Environmental Change Institute, University of Oxford, United Kingdom
Randall Bowie	Rockwool International A/S, United States
Rik Bogers	National Institute for Public Health and the Environment, the Netherlands
Jakob Bønløkke	National Institute for Public Health and the Environment, the Netherlands
Matthias Braubach	WHO European Centre for Environment and Health
Mauro Brolis	Finlombarda, S.p.A., Italy
Karl-Michael Brunner	*Wirtschaftsuniversität Wien* (Vienna University of Economics and Business Administration), Austria
Kathleen Cooper	Canadian Environmental Law Association (CELA), Canada
Eva Csobod	Regional Environmental Center for Central and Eastern Europe
Richard Cowart	Regulatory Assistance Project (RAP), Belgium
Anna Cronin de Chavez	Sheffield Hallum University, United Kingdom
Didier Cherel	*Agence de l'environnement et de la maîtrise de l'énergie (Ademe)* (Agency for Environment and Energy Management), government of France
Michael Davies	University College London (UCL), United Kingdom
Carlos Dora	Public Health and Environment, World Health Organization (WHO)
Ute Dubois	ISG Business School, France
Susanne Dyrboel	Rockwool International, A/S, United States
Søren Dyck-Madsen	The Ecological Council, Denmark
Odile Estibals	*Électricité de France* (EDF)
Mark Ewins	Department of Energy and Climate Change, government of the United Kingdom
Veronique Ezratty	*Service des Etudes Médicales* (Medical Studies Service), *Électricité de France* (EDF)
Deborah Frank	Boston Medical Center, United States
Nicolas Galiotto	Aalborg University, Denmark
Jan Gilbertson	Sheffield Hallam University, United Kingdom
Geoff Green	Sheffield Hallam University, United Kingdom
Ian Hamilton	UCL, United Kingdom
Marie-Eve Heroux	WHO
John Hills	London School of Economics, United Kingdom
Catherine Homer	Rotherham University, United Kingdom
Eva Hoos	Directorate-General for Energy, European Commission

© OECD/IEA, 2014.

Philippa Howden-Chapman	University of Otago, New Zealand
Victoria Johnson	Brotherhood of St Laurence, Australia
Dorota Jarosinska	European Environment Agency (EEA)
Matti Jantunen	National Institute of Health and Welfare, Finland
Zoltan Kapros	Hungarian Energy Office, government of Hungary
Ade Kearns	University of Glasgow, United Kingdom
Stylianos Kephalopoulos	DG Joint Research Centre, European Commission
Frederik Krogsøe	Ministry of Climate, Energy and Building, government of Denmark
Wilhelm Kuckshinrichs	*Forschungszentrum Juelich*, GmbH
Alaitz Landaluze	Innobasque, Spain
Jens Laustsen	Global Buildings Performance Network (GBPN)
Christine Liddell	University of Ulster, Northern Ireland
Marie Louyot-Gallicher	Research and Development, EDF
Rob Maas	National Institute for Public Health and the Environment, the Netherlands
Theo MacGregor	MacGregor Energy Consultancy, United States
Bettina Menne	Regional Office for Europe, WHO
Niamh McDonald	GBPN
Alex Macmillan	UCL, United Kingdom
Gergana Miladinova	DG Energy, European Commission
Helen McAvoy	Institute of Public Health in Ireland
Jacqueline McGlade	European Environment Agency (EEA)
Harald Meyer	Rockwool International, A/S, United States
Anca Moldoveanu	National Institute of Public Health, Hungary
Simon Nicol	Building Research Establishment (BRE), United Kingdom
Jerrold Oppenheim	Democracy and Regulation, Inc.
David Ormandy	Warwick Medical School, University of Warwick, United Kingdom
Charles O'Roarty	National Energy Action, Ltd., United Kingdom
Iben Østergaard	*Danish Teknologisk Institut*, Denmark
Christine Patterson	Energy Efficiency and Conservation Authority, government of New Zealand
Marianne Rappolder	Federal Environment Agency, government of Germany
Erin Rose	Oak Ridge National Laboratory (ORNL), United States
Jim Scheer	Sustainable Energy Authority of Ireland (SEAI)
Torben Sigsgaard	WHO
Damien Sullivan	Brotherhood of St Laurence, Australia
Ignacio Sánchez Díaz	United Nations Population Fund (UNFPA)
Hilary Thomson	University of Glasgow, Scotland
Angela Tod	Sheffield Hallum University, United Kingdom
Ellen Tohn	Tohn Environmental Strategies, United States

© OECD/IEA, 2014.

Agnieszka Tomaszewska	Institute for Sustainable Development, Poland
Bruce Tonn	Three3
Martin van den Berg	WHO
Helen Viggers	University of Otago, New Zealand
Ian Watson	Liverpool City Council, government of the United Kingdom
Jan Wammen	Dutch Energy Agency, government of the Netherlands
Volker Welter	United Nations Development Programme (UNDP)
Paul Wilkinson	London School of Hygiene and Tropical Medicine, United Kingdom

Industrial Productivity

Christiaan Abeelen	Netherlands Enterprise Agency (RVO), the Netherlands
Smail Alhilali	United Nations Industrial Development Organization (UNIDO)
Evelyn Bisson	Government of France
Mathieu Bordigoni	Research and Development, *Électricité de France* (EDF)
Nils Borg	European Council for an Energy Efficient Economy (ECEEE)
Randall Bowie	Rockwool International, A/S, United States
Didier Bosseboeuf	*Agence de l'environnement et de la maîtrise de l'énergie* (Ademe) (Agency for environment and energy management), government of France
Lars Brømsøe Termansen	Danish Energy Agency, government of Denmark
Conrad Brunner	Electrical Motor Systems Annex (EMSA), Switzerland
Bruno Chretien	Ademe, government of France,
Catherine Cooremans	University of Geneva, Switzerland
Patrick Crittenden	Sustainable Business, Pty. Ltd.
Maja Dahlgren	University of Linköping, Sweden
Annie Degen	United Nations Environment Programme (UNEP)
Greg Divall	Department of Industry, government of Australia
Bettina Dorendorf	Directorate-General for Energy (DG Energy), European Commission
Kristen Dyhr-Mikkelsen	Ea Energy Analyses
Steven Fawkes	EnergyPro Ltd., United Kingdom
Hugh Falkner	Electric Motor Systems Annex (EMSA), Switzerland
Pedro Faria	Carbon Disclosure Project, United Kingdom
Erik Gudbjerg	Lokalenergi, Denmark
Andreas Guertler	European Industrial Insulation Foundation (EIIF), Switzerland
William Garcia	European Chemical Industry Council (CEFIC)
Ali Hasanbeigi	Lawrence Berkeley National Laboratory (LBNL), United States
Michiel Hekkenberg	Energy Research Centre of the Netherlands (ECN), the Netherlands
Rod Janssen	Consultant

© OECD/IEA, 2014.

Etienne Kechichian	International Finance Corporation (IFC), World Bank
Osamu Kimura	Central Research Institute of Electric Power Industry, Japan
Mirko Krueck	Local Energy Efficiency Networks (LEEN), Germany
Ruben Kubiak	DG Energy, European Commission
Sarah Larsen	Danish Energy Agency, government of Denmark
Ian Leslie	Leslie Consulting
Robert Bruce Lung	Consultant
Sarah Meagher	Department of Energy and Climate Change, government of the United Kingdom
Luis Mundaca	Lund University, Sweden
Jean-Jacques Marchais	Schneider Electric
Hannes MacNulty	Consultant
Darcee Meilbeck	Organisation for Economic Co-operation and Development (OECD)
Aimee McKane	Lawrence Berkeley National Laboratory (LBNL), United States
Evan Mills	LBNL, United States
Therese Nehler	University of Linköping, Sweden
Andi Novianto	Ministry for Economic Affairs, government of Indonesia
John O'Sullivan	Sustainable Energy Authority of Ireland
Alexios Pantelias	IFC, World Bank
Lynn Price	LBNL, United States
Josephine Rasmussen	University of Linköping, Sweden
Julia Reinaud	Institute for Industrial Productivity (IIP)
Clemens Rodhe	Institute for Systems and Innovation Research (ISI), Germany
Fredrich Seefeldt	Prognos, Germany
Jigar Shah	IIP
Anjan Sinha	National Productivity Council, India
Jonathan Sinton	World Bank
Johan Slobbe	ECN, the Netherlands
Christian Stenqvist	Lunds University, Sweden
Peter Sweatman	Climate Strategy
Andrea Trianni	*Politecnico Di Milano* (Milan Polytechnic), Italy
Maartin Van Werkhoven	Consultant
Hannu Vaananen	ABB, Finland
Ton Van Dril	ECN, the Netherlands
Maarten Van Werkhoven	Consultant
Ernst Worrell	Copernicus Institute of Sustainable Development, the Netherlands
Eric Woodroof	Profitable Green Solutions, United States

© OECD/IEA, 2014.

Energy delivery

Nicholas Abi-Samra	DNV KEMA, France
Karl Abraham	Environment Canada, government of Canada
Riley Allen	Regulatory Assistance Project (RAP)
John Appleby	Natural Resources Canada (NRCan), government of Canada
Elham Azarafshar	NRCan, government of Canada
Darcy Blais	NRCan, government of Canada
Francis Bradley	Canadian Electricity Association, Canada
Justin Brant	Massachusetts Department of Public Utilities, United States
Maxine Bretzlaff	Canadian Gas Association, Canada
Carol Buckley	NRCan, government of Canada
Jim Burpee	Canadian Electricity Association, Canada
Vicki Campbell	DTE Energy, United States
Paul Cheliak	Canadian Gas Association, Canada
Yanick Clément-Godbout	NRCan, government of Canada
Rebecca Craft	Con Edison, United States
Ken Colburn	RAP
Michele Coughlan	Newfoundland Power, Canada
Benjamin Davis	Massachusetts Department of Public Utilities, United States
Jamie Drakos	The Cadmus Group, United States
Tom Eckman	Northwest Power Planning Council, United States
Tim Egan	Canadian Gas Association, Canada
Chuck Farmer	Ontario Power Authority, Canada
Janice Garcia	Canadian Electricity Association, Canada
Howard Geller	Southwest Energy Efficiency Project, United States
Mimi Goldberg	DNV KEMA
Bryan Gormley	Canadian Gas Association, Canada
Chuck Gray	National Association of Regulatory Utility Commissioners, United States
Aaron Hoskin	NRCan, government of Canada
Doug Hurley	Synapse Energy Economics, Inc., United States
Jennifer Kallay	Synapse Energy Economics, Inc., United States
Ann Kelly	Canadian Electricity Association, Canada
Jay Khosla	NRCan, government of Canada
Matthew Lam	NRCan, government of Canada
Krista Langthorne	Newfoundland Power, Canada
Jack Laverty	Columbia Gas, United States
Claude Lefrançois	NRCan, government of Canada
Adi Lyer	Market Development Canadian Gas Association, Canada
Joy Morgenstern	California Public Utilities Commission, United States

© OECD/IEA, 2014.

Émilie Moorhouse	Gaz Métro, Canada
Steve Nadel	American Council for an Energy-Efficient Economy (ACEEE)
Laura Oleson	NRCan, government of Canada
Brandon Ott	DSM Policy Enbridge Gas Distribution Inc., Canada
Mitch Rosenberg	KEMA Inc., the Netherlands
Giuliana Rossini	Hydro One, Canada
Shahana Samiullah	Southern California Edison, United States
Rich Sedano	RAP
Lisa Skumatz	Skumatz Economic Research Associates, Inc., United States
Kelly Smith	NRCan, government of Canada
Sarah Smith	FortisBC, Canada
Elaine Prause	Energy Trust Oregon, United States
Marie Lyne Tremblay	NRCan, government of Canada
Bruce Tonn	Oak Ridge National Laboratory (ORNL), United States
Chris Tyrrell	Toronto Hydro, Canada
Laura van Wie	Alliance to Save Energy, United States
Eric Winkler	Independent System Operator (ISO) New England, United States
Ed Wisniewsk	Consortium for Energy Efficiency, United States
Lisa Wood	Institute for Electric Efficiency, United States
Tim Woolf	Synapse Economics Inc., United States

This work was supported by financial and in-kind contributions from several IEA member countries and external organisations:

- Department of Industry, government of Australia
- Natural Resources Canada, government of Canada
- Sustainable Energy Authority of Ireland
- Department of Energy and Climate Change, government of the United Kingdom
- Climate Works
- European Environment Agency
- European Copper Institute

© OECD/IEA, 2014.

Executive Summary

As energy efficiency continues to gain attention as a key resource for economic and social development across all economies, understanding its real value is increasingly important. The multiple benefits approach to energy efficiency policy seeks to expand the perspective of energy efficiency beyond the traditional measures of reduced energy demand and lower greenhouse gas (GHG) emissions by identifying and measuring its impacts across many different spheres.

The term "multiple benefits"[1] aims to capture a reality that is often overlooked: investment in energy efficiency can provide many different benefits to many different stakeholders. Whether by directly reducing energy demand and associated costs (which can enable investment in other goods and services) or facilitating the achievement of other objectives (e.g. making indoor environments healthier or boosting industrial productivity), recent research acknowledges the enormous potential of energy efficiency. This publication demonstrates its role as a major contributor to strategic objectives across five main themes: enhancing the sustainability of the energy system, economic development, social development, environmental sustainability and increasing prosperity.

Underpinned by a comprehensive review of existing evidence, the aim of this book is two-fold: to build knowledge and understanding of the nature and scope of the multiple benefits of energy efficiency, and to provide practical guidance on how to apply policy development and assessment tools to account for these impacts. The combination of theory and practice will help policy makers and other stakeholders to integrate multiple benefits into strategic planning in order to maximise the potential for positive outcomes.

Energy efficiency: The "first fuel" with large untapped potential

Energy efficiency is taking its place as a major energy resource in the context of national and international efforts to achieve sustainability targets. This reflects a paradigm shift that is beginning to give credence to actions on both the supply and the demand side in the quest to achieve economic growth while supporting energy security, competitiveness and environmental sustainability.

In effect, attention to energy efficiency has begun to evolve, progressing from the lack of visibility inherent in its identification as "the hidden fuel" (i.e. measured and valued only as the negative quantity of energy not used) to an increasing recognition of its role as the "first fuel". Energy use avoided by International Energy Agency (IEA) member countries in 2010 (generated from investments over the preceding 1974 to 2010 period), was larger than actual demand met by any other single supply-side resource, including oil, gas, coal and electricity – making energy efficiency the largest or "first" fuel. Aggregate annual investments in energy efficiency have been estimated at USD 300 billion in 2011, which is equal to aggregate investments in coal, oil and gas power generation. Macroeconomists have stated that energy efficiency is the surest energy supply that exists. Harnessing economically viable energy efficiency investments would facilitate a more efficient

1 In other literature, these impacts have been variously labelled "co-benefits", "ancillary benefits" and "non-energy benefits" – terms often used interchangeably with "multiple benefits". The IEA uses the term multiple benefits, which is broad enough to reflect the heterogeneous nature of outcomes and to avoid pre-emptive prioritisation of various benefits; different benefits will be of interest to different stakeholders.

© OECD/IEA, 2014.

allocation of resources across the global economy, with the potential to boost cumulative economic output through 2035 by USD 18 trillion – larger than the current size of the economies of North America combined (namely, the United States, Canada and Mexico). Energy efficiency has also become a pillar of global development goals, including the United Nations Sustainable Energy for All initiative. In the face of rising energy demand, global growth aspirations and the pressing need to limit GHG emissions, the market for energy efficiency could develop rapidly – provided that stakeholders understand its value.

Notwithstanding this emerging role for energy efficiency, future projections reveal that under existing policies, the vast majority of economically viable energy efficiency investments will remain unrealised (Figure ES.1).

Figure ES.1 Long-term energy efficiency economic potential by sector

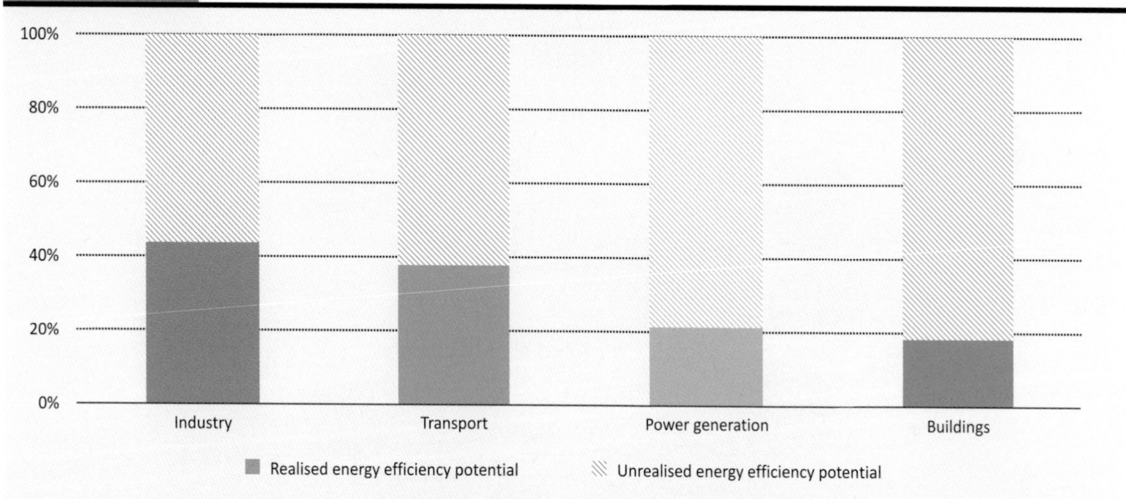

Note: These energy efficiency potentials are based on the IEA New Policies Scenario outlined in the World Energy Outlook 2012. Investments are classified as "economically viable" if the payback period for the up-front investment is equal to or less than the amount of time an investor might be reasonably willing to wait to recover the cost, using the value of undiscounted fuel savings as a metric. The payback periods used were in some cases longer than current averages but they were always shorter than the technical lifetime of individual assets.
Source: IEA (2012), *World Energy Outlook 2012*, OECD/IEA, Paris.

Key point IEA projections to 2035 show that as much as two-thirds of energy efficiency potential will remain untapped unless policies change.

Many barriers contribute to the limited uptake of energy efficiency opportunities;[2] one main obstacle is the lack of attention paid to energy efficiency investment opportunities by stakeholders in both the private and government sectors relative to supply-side opportunities, including new resources such as shale gas and oil. The multiple benefits approach seeks, in part, to address this barrier by rendering more apparent the benefits that energy efficiency can generate for these stakeholders. It also helps to address the challenge of the invisibility of energy efficiency (i.e. representing energy *not* used), by appropriately crediting it with the value of the positive impacts it triggers across a variety of areas.

2 These include information failures, split incentives, subsidised pricing of energy, inadequate pricing of externalities and a shortage of financing.

© OECD/IEA, 2014.

Capturing the multiple benefits of energy efficiency

Research has brought to the fore a range of areas, beyond energy demand reduction and lower GHG emissions, in which clear benefits of energy efficiency have been documented (Figure ES.2). Most of these benefits are relevant to IEA member countries and non-member countries alike, although prioritisation by individual countries is likely to vary. Experts increasingly acknowledge the important role of energy efficiency in generating a broad range of outcomes that support ambitions to improve wealth and welfare – goals that the public and policy makers both understand and aspire to achieve.

Figure ES.2 The multiple benefits of energy efficiency improvements

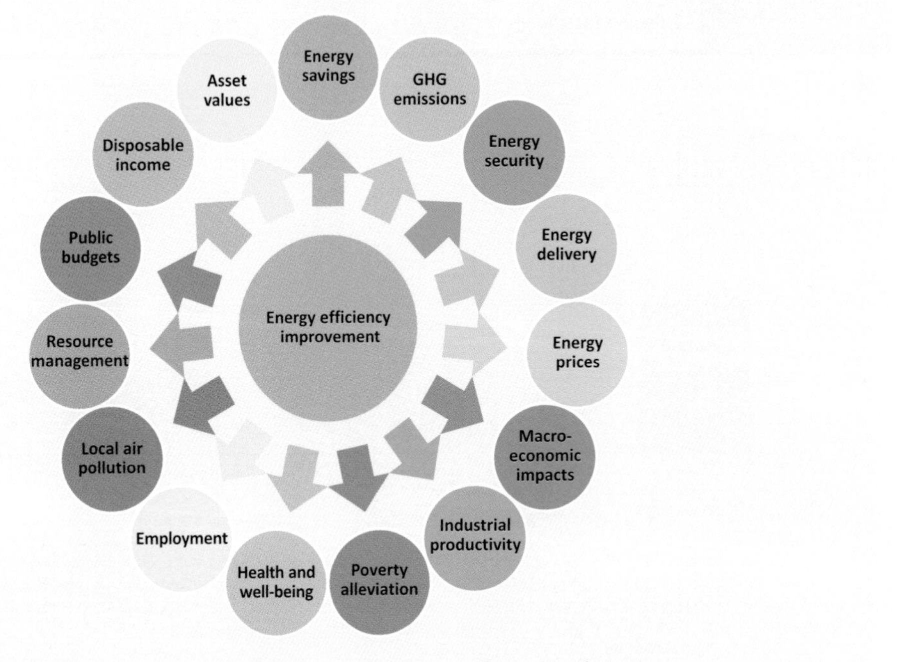

Note: This list is not exhaustive, but represents some of the most prominent benefits of energy efficiency identified to date.
Source: Unless otherwise noted, all material in figures and tables in this chapter derives from IEA data and analysis.

Key point A multiple benefits approach to energy efficiency reveals a broad range of potential positive impacts.

To date, these broader impacts of energy efficiency have not been systematically assessed, in part due to a lack of critical data and the absence of mature methodologies to measure their scope and scale. As a result, the degree to which energy efficiency could enhance economic and social development is not well understood, and generally considered in national policy decision-making processes only in a qualitative way, if at all.

In examining how methodologies – many of which are well known to economic and policy evaluation – can be applied to the multiple benefits of energy efficiency, this publication demonstrates how often overlooked, and even intangible, outcomes can be captured, offering the possibility to send better socio-economic signals to complement market signals. Strengthening capacity in both the public and private sectors to better assess the full range of outcomes of energy efficiency will improve both the basis for economic analysis of policy options and the ability to communicate the value that energy efficiency can deliver for the economy and society.

© OECD/IEA, 2014.

A deep dive into key benefit areas

Drawing on best available information from governments and academia, this book provides in-depth analysis of five benefit areas: macroeconomic development; public budgets; health and well-being; industrial productivity; and energy delivery. These areas were chosen for two reasons:

- they tend to be policy priorities in IEA member countries and beyond
- enough evidence is available about their potential impacts to begin robust analysis.

Macroeconomic development

Macroeconomic assessment is a mainstream branch of economic analysis that has built up a huge body of knowledge and evidence over many years; however, the impact of energy efficiency policies on macroeconomic performance still needs to be better understood and systematically measured. Energy efficiency improvements can deliver benefits across the whole economy, with direct and indirect impacts on economic activity (measured through gross domestic product [GDP]), employment, trade balance and energy prices. In general, analysis of GDP changes due to large-scale energy efficiency policies show positive outcomes with economic growth ranging from 0.25% to 1.1% per year. The potential for job creation ranges from 8 to 27 job years per EUR 1 million invested in energy efficiency measures. How energy efficiency measures influence these areas (i.e. positively or negatively) depends on a country's economic structure and on the design and scale of the underlying policies.

Public budgets

Whether by reducing government expenditures on energy or by generating increased tax revenues through greater economic activity and/or increased spending on energy efficiency-related and other goods and services, energy efficiency improvements can have important impacts on the budgetary position of national and sub-sovereign entities. An important impact on public budget is on reduced fuel costs for heating, cooling and lighting, a budget line that is expected to increase over time as energy prices rise. One of the greatest impacts overall is the reduced budget for unemployment payments when energy efficiency policies lead to job creation. Public budget impacts are thus closely linked to macroeconomic impacts.

Although most governments have developed methodologies to estimate the costs and benefits of a policy to the public budget, the full range of public budget benefits are rarely estimated. This broader range of benefits can multiply the calculated value to the public budget by two or three times. An initial evaluation of initiatives to advance energy efficiency in buildings, for example, calculated a value of USD 41 billion to USD 55 billion (EUR 30 billion to EUR 40 billion) to the European public budget; adding tax revenues and reduced unemployment payments increased the value to USD 91 billion to USD 175 billion (EUR 67 billion to EUR 128 billion). Similarly, reduced energy demand can create long-term, cumulative savings for governments that subsidise energy production and consumption.

Health and well-being

Energy efficiency retrofits in buildings (e.g. insulation retrofits and weatherisation programmes) create conditions that support improved occupant health and well-being, particularly among vulnerable groups such as children, the elderly and those with pre-existing illnesses. The potential benefits include improved physical health such as reduced symptoms of respiratory and cardiovascular conditions, rheumatism, arthritis and allergies, as well as fewer injuries. Several studies that quantified total outcomes found benefit-cost ratios as high as 4:1 when health and well-being impacts were included, with health

© OECD/IEA, 2014.

benefits representing up to 75% of overall benefits. Improved mental health (reduced chronic stress and depression) has, in some cases, been seen to represent as much as half of total health benefits.

Realised health improvements generate downstream social and economic impacts, including lower public health spending. Addressing indoor air quality through energy efficiency measures could, in a high energy efficiency scenario, save the European Union's economy as much as USD 259 billion (EUR 190 billion) annually.

Industrial productivity

Industry often views energy as an operational cost; energy savings are perceived as incidental benefits of other investments rather than as a central value-generating proposition. Yet, industrial energy efficiency measures deliver substantial benefits in addition to energy cost savings – enhancing competitiveness, profitability, production and product quality, and improving the working environment while also reducing costs for operation and maintenance, and for environmental compliance. Introducing multiple benefits can help to better align energy efficiency with strategic business priorities, thereby strengthening the business case for investment. The value of the productivity and operational benefits derived can be up to 2.5 times (250%) the value of energy savings (depending on the value and context of the investment). Including such productivity outcomes in financial cost assessment frameworks can substantially reduce the payback period for energy efficiency investment, in some cases from four years to one year.

Energy delivery

Even utilities and other energy providers gain in a variety of ways from energy efficiency measures. Direct benefits include lower costs for energy generation, transmission and distribution, improved system reliability, dampened price volatility in wholesale markets and the possibility of delaying or deferring costly system upgrades. Providers can also benefit indirectly through benefits that accrue to customers from improved affordability of energy services, which in turn can reduce arrears and the associated administrative costs for utilities. To date, these and other customer benefits have proven difficult to integrate properly into cost-effectiveness tests and therefore have not been accurately measured; however, standard practice valuation frameworks are being developed to accommodate measurement of a broader range of benefits for energy providers and their customers.

Different country perspectives on benefits

While all of the benefits described above are likely to be relevant in all economies, national circumstances, as well as economic and social priorities, will play important roles in their prioritisation. Different countries will value distinct benefits differently, and within a specific national context, different stakeholders will be interested by different benefits. Many developing countries with low energy access rates, for example, can use energy efficiency to service more customers from a given asset base. In countries with near universal access rates, improved industrial productivity may be the main driver for energy efficiency. Given projections of substantial economic growth and related energy demand increase in emerging and developing economies, multiple benefits analysis can and should be adapted to their specific needs and challenges (Box ES.1).

A few cautions to keep in mind

The effects that drive these benefits are dynamic and present several complexities that are important to consider when applying a multiple benefits approach to assessment of energy efficiency policies.

© OECD/IEA, 2014.

Box ES.1

Energy efficiency generates important benefits for emerging economies

Improved energy efficiency provides a variety of benefits of particular importance for emerging economies and developing countries as they seek to exploit their resource base to reduce poverty and support sustainable growth:

- **Access**: Energy efficiency can help countries to expand access, effectively enabling them to supply power to more people through the existing energy infrastructure.
- **Development/growth**: Energy efficiency has a variety of positive impacts that support economic growth, for example by improving industrial productivity and reducing fuel import bills.
- **Affordability/poverty alleviation**: Energy efficiency can increase the affordability of energy services for poorer families by reducing the per-unit cost of lighting, heating, refrigeration and other services.
- **Local pollution**: Energy efficiency (both supply side and end-use) can help to reduce the need for generation – and lower associated emissions – while supporting economic growth.
- **Climate change resilience**: By reducing the need for energy infrastructure, energy efficiency reduces the amount of energy assets exposed to extreme weather events.

- A holistic and comprehensive analytical approach is needed to enable evaluators to assess whether individual effects – and also the net effect – of a particular energy efficiency policy is positive or negative.
- Some benefits are indirect or are the product of a chain of effects; identifying a causal link between a particular energy efficiency measure and a specific benefit can present a challenge and requires analytical effort.
- Impacts can occur simultaneously at different levels of the economy – from individual citizens or households or sub-sectoral to sectoral levels – at the national or international scale. They can also create flow-on impacts, making it extremely important to establish mechanisms to avoid double-counting of benefits when comparing benefits with costs.

The rebound effect

One of the most persistent challenges in energy efficiency policy is accounting for the phenomenon known as the "rebound effect" – where improved energy efficiency is used to access more energy services rather than to achieve energy demand reduction. A multiple benefits perspective helps to understand the impacts, as well as the sources and causes, of an observed rebound effect and accordingly to manage better any trade-offs that might take place. Some benefits can come with an energy consumption price tag (e.g. when improved energy affordability leads to increased consumption of heating). Where energy savings are "taken back" in the achievement of health benefits, poverty alleviation, or improving productivity, the rebound effect can be viewed as having a net positive outcome, amplifying the benefits of the energy efficiency intervention. In some contexts, such as developing countries with high growth rates where activity tends to be more energy-intensive, rebound may often be desirable as it enables the economy to capitalise further on its energy resources and stimulate other efficiencies.

It remains important to fully assess any potential rebound effects, taking them into account when calculating the actual energy demand reductions (in particular when the objective is reduced pollution, reduced fuel imports, or GHG emissions reductions tied to lower electricity generation). One must also consider the rebound effect against the backdrop of

© OECD/IEA, 2014.

the particular energy efficiency policy, the specific benefit(s) being targeted and the relevant economic conditions. Unbundling the relationship between energy savings and the broader outcomes of energy efficiency can provide a fuller understanding of the rebound effect, and a clearer appreciation of where this effect either reduces or amplifies the benefits of an energy efficiency intervention.

Putting the tools in the hands of policy makers

Thorough evaluation of the impacts of energy efficiency policies across a variety of areas underpins the multiple benefits approach. To provide policy makers with better information to develop and evaluate energy efficiency policies (and broader energy policy portfolios), the assessment must go beyond merely measuring energy demand reductions.

Analysis of multiple benefits needs to be supported by a robust evidence base. This requires finding better ways to measure, quantify and, ideally, monetise benefits so that they can be integrated into existing policy assessment frameworks. Most governmental policy assessment guidelines[3] recommend consideration of a range of social, economic and environmental issues. In practice, multiple benefits assessment remains limited for two reasons: methods for assessing the costs and benefits of non-market impacts have not been fully developed; and such assessments require greater resources (financial and human) than more traditional policy evaluation methods.

Of the range of existing tools and methods governments use for policy assessment, some are better suited than others to measuring multiple benefits. The choice of tools depends on several factors, including the time and resources available for the analysis, and the quality of available data. However, even rough estimates provide more accurate information on which to base policy decisions than assuming a value of zero for observed benefits. Experts advocate for the triangulation of evidence from mixed methodologies to ensure that all impacts reported by beneficiaries of an energy efficiency intervention are taken into account and verified when assessing the net value of an energy efficiency policy.

This book demonstrates how currently available tools can be used to put the multiple benefits approach into practice. Existing policy assessment tools are expected to develop and new ones will emerge to better serve the growing interest of policy makers in the impacts of energy efficiency, and support their ability to optimise energy efficiency policies.

Conclusion

Applying a multiple benefits approach to energy efficiency policy should enable a fuller understanding of the potential of energy efficiency. It signals a shift away from the traditional view of energy efficiency as simply delivering energy demand reductions, and recognises its important role in delivering concrete social and economic improvements. This shift could initiate a step-change in the uptake of energy efficiency opportunities and spur the international community onto an economically efficient path to achieving shared development goals. The multiple benefits approach could accelerate the shift of energy efficiency from its status as the "hidden fuel" to its emerging role as the "first fuel".

This book aims to articulate the scope and scale of several of these benefits, thereby building the case for considering multiple benefits as a matter of course in the energy efficiency policy process. A cross-section of emerging evidence provides evaluators with examples of how to better account for the benefits and costs of energy efficiency measures to support design and implementation of policy portfolios that maximise

3 For example, the *Green Book* in the United Kingdom (UK HM Treasury, 2003), the *Impact Assessment Guidelines* issued by the European Commission (EC, 2009) and the *California Standard Practice Manual* which is used throughout the United States (CPUC, 2001).

© OECD/IEA, 2014.

prioritised benefits. Ultimately, improved knowledge will enable governments to make their own assessment of the value of the multiple benefits approach, in line with national strategies. Significant further work is needed to deepen understanding of the policy-outcome dynamics at work, to improve metrics for measuring multiple benefits, and to continue building the evidence base to support policy decision making.

Each country taking steps towards a multiple benefits approach will have unique priorities, and will need to adapt the approach to its particular national context. This will influence evaluation requirements and data resources needed. Armed with more comprehensive information about the value of energy efficiency, countries will be better able to design energy efficiency policies that maximise these additional positive impacts within the context of achieving prioritised policy objectives.

Bibliography

CPUC (California Public Utilities Commission) (2001), *California Standard Practice Manual Economic Analysis of Demand-Side Programs and Projects*, Governor's Office of Planning and Research, State of California, Sacramento, http://cleanefficientenergy.org/sites/default/files/07-J_CPUC_STANDARD_PRACTICE_MANUAL.pdf (accessed 28 June 2014).

EC (European Commission) (2009), *Impact Assessment Guidelines*, Sec(2009) 92, EC, Brussels, http://ec.europa.eu/smart-regulation/impact/commission_guidelines/docs/iag_2009_en.pdf (accessed 22 June 2014).

IEA (2013), *Energy Efficiency Market Report 2013*, OECD/IEA, Paris.

IEA (2012), *World Energy Outlook 2012*, OECD/IEA, Paris.

UK HM Treasury (United Kingdom, Her Majesty's Treasury) (2003), *The Green Book: Appraisal and Evaluation in Central Government*, United Kingdom government, London.

© OECD/IEA, 2014.

Chapter 1

Taking a multiple benefits approach to energy efficiency

Key points

- *A growing body of evidence shows that energy efficiency can deliver substantial value through a broad range of economic and social impacts beyond the traditional focus on energy demand reduction. This publication uses the term the "multiple benefits" of energy efficiency to encompass a broad range of positive impacts across diverse sectors.*
- *Assessments show that the scale of the multiple benefits can be substantial: some impacts of improved energy efficiency delivered as much as 2.5 times the value of the energy demand reduction. Broadly, energy efficiency can stimulate economic and social development, enhance energy system sustainability, contribute to environmental sustainability and increase prosperity.*
- *Efforts to identify and quantify these multiple benefits can provide the data needed to inform sound policy development and associated resource allocation decisions, in large part by facilitating more accurate assessment of the trade-offs between reduced energy demand and socio-economic welfare gains of a particular energy efficiency policy.*
- *Energy cost savings from energy efficiency measures can stimulate a "rebound effect", in which expected energy demand reductions are not achieved because financial savings from lower energy costs are reinvested in more goods and services. The multiple benefits approach reveals that the rebound effect can be a positive thing, if the goods and services acquired are of greater value to society.*
- *Widespread adoption of the multiple benefits approach could stimulate higher uptake of energy efficiency programmes and measures, ultimately shifting energy efficiency from "the hidden fuel" to an increasing recognition of its role as the "first fuel" among energy resources.*

Introducing the multiple benefits of energy efficiency

Energy is the fundamental fuel for economic and social development, and energy efficiency measures boost development, by increasing the amount of "service" gained from every unit of energy. A major driver behind energy efficiency investments is its capacity to lower energy demand and deliver energy cost savings.

In recent years, more attention has been paid to the notion that energy efficiency helps to achieve a much broader range of outcomes that contribute to the human ambition to improve welfare and wealth. These benefits include various macroeconomic benefits (e.g. shifts in energy trade balances and employment), increased access to energy and

© OECD/IEA, 2014.

improved affordability of energy services, reduced air pollution, and fiscal improvements for national and sub-national entities. The International Energy Agency (IEA) refers to this suite of outcomes as the "multiple benefits"[1] of energy efficiency (Figure 1.1).

Figure 1.1 The multiple benefits of energy efficiency

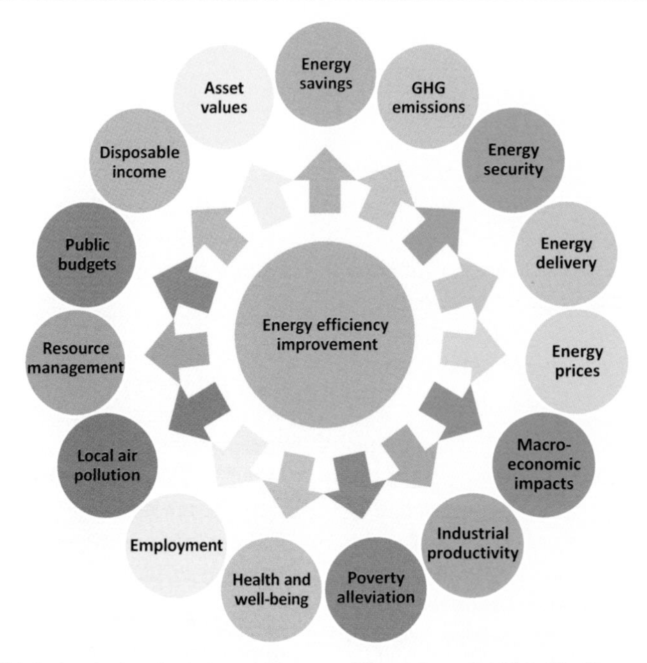

Notes: GHG = greenhouse gas. This list is not exhaustive, but represents some of the most prominent benefits of energy efficiency identified to date.
Source: Unless otherwise noted, all material in figures and tables in this chapter derives from IEA data and analysis.

Key point *A multiple benefits approach to energy efficiency reveals a broad range of potential positive impacts.*

Of these benefits, only reductions in energy demand and GHG emissions have been measured systematically to date. Yet a growing body of evidence demonstrates that the broader benefits may be extremely valuable, in both economic and social terms. The multiple benefits of energy efficiency extend to goals that the general public understand and may personally aspire to. They are, therefore, also of great interest to policy makers. A positive effect in any one of these areas may generate at least equal (if not even greater) interest for the public and for policy makers than the energy savings achieved. Such added value offers a powerful economic and social signal with the potential to motivate energy efficiency action.

The degree to which optimal energy efficiency could enhance economic and social development is not sufficiently well understood, and generally is considered in national policy decision-making processes only in a qualitative way, if at all. While energy efficiency experts and many policy makers are alert to the fact that energy efficiency generates

1 In other literature, these impacts have been variously labelled "co-benefits", "ancillary benefits" and "non-energy benefits" (NEBs) – and are often used interchangeably with "multiple benefits". The IEA uses the term multiple benefits, which is broad enough to reflect the heterogeneous nature of outcomes and to avoid pre-emptive prioritisation of various benefits; different benefits will be of interest to different stakeholders.

© OECD/IEA, 2014.

broader impacts (particularly in the areas of employment and economic growth), these impacts have not been systematically assessed.[2] Much of the critical data on broader impacts are found outside the energy sector. What figures are available often derive from modelling and scenario analyses.[3] While such analysis provides useful insights, it is a departure from standard approaches and reflects the absence of sufficiently mature methodologies to measure the scope and scale of broader impacts of energy efficiency.

Recognising the challenge of accurate assessment, this publication seeks to identify and promote the development of more robust methods for measuring multiple benefits so that these important social and economic effects may be better integrated and evaluated to better inform the policy decision-making process.

Energy efficiency: From "hidden fuel" to "first fuel"

Energy efficiency is usually defined by the ratio of energy consumed to the output produced or service performed.

Reducing energy demand, or conserving energy, has been the main driver for energy efficiency policies in many countries. This goal is pursued primarily by improving the efficiency of energy-consuming products and processes on both the demand side and the supply side of the energy equation. Energy efficiency policies remain one of the most effective tools for achieving energy conservation goals.

A portfolio of measures can improve the efficiency of energy-using goods, services and processes across all sectors. This portfolio includes policies that target the demand side, for example establishing building energy codes, setting minimum energy performance requirements for energy-using equipment, setting standards to improve vehicle fuel efficiency, developing efficient public transport systems and promoting energy management systems in industry. Other energy efficiency measures focus on improving efficiency in energy supply, for example by reducing technical losses in generation, transmission and distribution. The IEA *25 Energy Efficiency Policy Recommendations* highlight the importance of ensuring that energy efficiency policies are well-designed and connected to realistic targets that are accurately monitored.

The traditional view of energy efficiency

Traditionally, the impact of energy efficiency policies has been measured purely in terms of units of reduced energy demand; i.e. the level of improvement in the ratio of energy to output, as measured over time against a baseline or business-as-usual (BAU) energy consumption scenario. Measuring a negative value – the energy not consumed or the energy costs avoided – can seem somewhat intangible and has led many commentators to refer to energy efficiency as the "hidden fuel". In the case of energy efficiency, substantial uncertainties about the behaviour of energy users come into play, complicated by the social and economic factors that influence how the efficiency improvement is taken up. Thorough impact assessment

2 Although current efforts are limited, some countries have made progress with exploring other impacts. Important studies on the employment impact of its energy efficiency programmes have been carried out by many IEA member countries individually and by international organisations such as the International Labor Organization (ILO), the Organisation for Economic Co-operation and Development (OECD) and the United Nations Environment Programme (UNEP) (see Chapter 2).

3 IEA assessments of the potential impact of energy efficiency investments systematically include two elements: forward projections of the impact on gross domestic product (GDP) growth of economically viable energy savings and the importance of these savings (over the medium to long term) for climate change mitigation.

© OECD/IEA, 2014.

can generally take account of all of these issues when robust methods for measuring energy demand reductions are applied (see the Companion Guide at the end of this publication).

A persistent challenge in energy efficiency policy is that results often fall short of the full energy savings projected when technical potentials are estimated during the policy development phase. Efficiency improvements that save energy, thus resulting in lower energy costs, can stimulate increased energy consumption and general expenditures that counteract the technical potential savings. This phenomenon, referred to as a "rebound effect", is usually perceived as a negative outcome. The multiple benefits approach has important implications for better understanding what the rebound effect represents in terms of welfare improvements for society, revealing that it can, in fact, generate additional benefits (see further discussion of the rebound effect in a multiple benefits context below).

In many countries, energy efficiency has emerged as a significant means of reducing GHG emissions to support climate change mitigation strategies. When energy efficiency measures result in energy demand reduction, in particular lower consumption of fossil fuels, an additional benefit is reduced GHG emissions (Box 1.1).

Box 1.1 Energy efficiency as a tool to reduce GHG emissions

Many governments now strategically align energy efficiency policies and GHG reduction goals. Once projected or actual energy savings have been established, these figures can be converted – on the basis of the carbon intensity of the avoided fuel or altered supply mix – to an equivalent value for GHG emission savings (measured as tonnes of carbon dioxide [CO_2] saved per unit of energy). Correct accounting for the rebound effect may reduce the potential contribution of energy efficiency to climate change mitigation, possibly altering the relative priority of different CO_2 abatement policies.

Scenario modelling carried out for the IEA *Energy Technology Perspectives 2014* show that energy efficiency measures can contribute about 40% of the CO_2 abatement needed by 2050 to achieve emissions reduction consistent with a target of limiting global temperature increase to 2 degrees Celsius.* When compared against other abatement options (nuclear, carbon capture and storage, and even renewable energy), energy efficiency offers substantial benefits in terms of cost-effectiveness and shorter lead times for delivery of GHG reductions.

* As decided by the Conference of the Parties to the United Nations Framework Convention on Climate Change (UNFCCC) in Copenhagen in 2009.

The emergence of energy efficiency as "the first fuel"

More recently, energy efficiency has come to be recognised as a major energy resource. Thanks to energy efficiency investments over the preceding 1974 to 2010 period, energy use avoided by IEA member countries in 2010 was larger than any other single supply-side resource, including oil, gas, coal and electricity. This reality supports the concept of recognising energy efficiency as the "first fuel" for many IEA member countries.

The uptake of economically viable energy efficiency investments (as posited in the IEA Efficient World Scenario [EWS]) would result in a more efficient allocation of resources across the global economy, with the potential to boost cumulative economic output through 2035 by USD 18 trillion – larger than the current size of the economies of North America combined (namely, Canada, Mexico and the United States). Additional investment of USD 11.8 trillion in more efficient end-use technologies is needed, but is more than

© OECD/IEA, 2014.

offset by a USD 17.5 trillion reduction in fuel expenditures and USD 5.9 trillion in lower supply-side investment.

In 2013, the IEA undertook an analysis of the market for energy efficiency, adding the first *Energy Efficiency Market Report* (IEA, 2013) to its series of annual reports on fuel markets. Analysis of investments in energy efficiency shows an annual global market for energy efficiency products and services worth USD 300 billion in 2011 (Figure 1.2).[4]

Figure 1.2 Global levels of investment in selected areas of the energy system in 2011

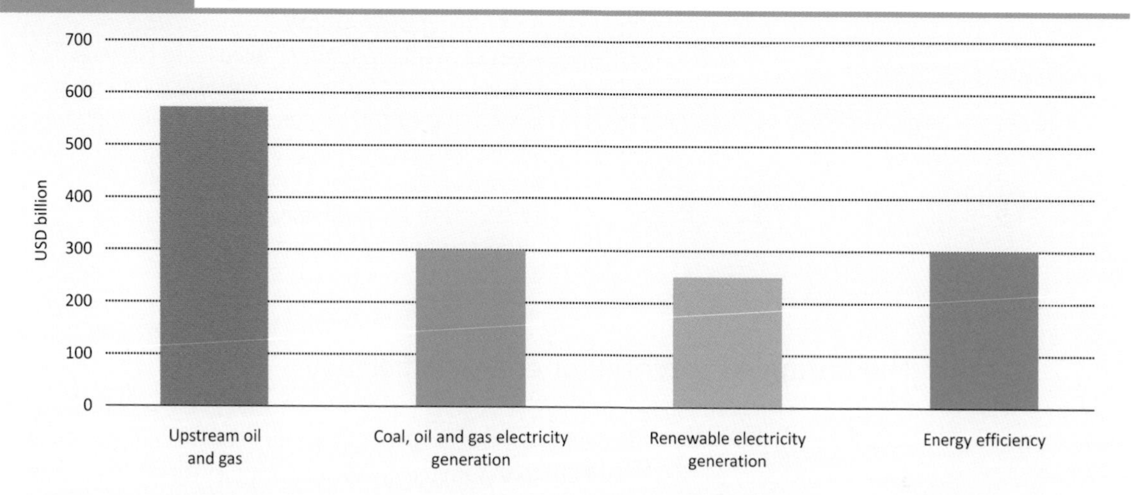

* Estimated range of USD 147 billion to USD 300 billion. Investment figures include public and private investment.
Source: IEA (2013), *Energy Efficiency Market Report 2013: Market Trends and Medium-Term Prospects*, OECD/IEA, Paris.

Key point *The scale of global energy efficiency market is comparable to supply-side investment markets.*

Attention to energy efficiency has been growing as part of national efforts to achieve sustainability targets. This is reflected, for example, in recent revisions of the European energy targets, which emphasise the role for energy efficiency measures in achieving various objectives within member states (EC, 2012). Energy efficiency is also playing an increasing role in achievement of global goals. Within the context of the UNFCCC, the negotiation process has recently focused on energy efficiency as a key measure with high potential to assist member countries in meeting climate change mitigation goals. Similarly, energy efficiency is identified as one of three pillars in the United Nations (UN) Sustainable Energy for All (SE4ALL) initiative, which promotes shared goals for sustainable development (Box 1.2). Numerous other international co-operative initiatives to promote energy efficiency underline the increasing recognition of its positive potential impact.

4 The finding is based on a country-by-country analysis of reported public spending, combined with information about multilateral institutional investment and private spending (where available). Country sources and estimates were used when available. When no data were available, multipliers (in the form of leverage ratios) were applied to data from multilateral development banks and other relevant sources of public funding to estimate the private finance leveraged by public funding of energy efficiency across the economy.

© OECD/IEA, 2014.

Box 1.2 Energy efficiency as a tool for development in the UN SE4ALL initiative

In September 2011, UN Secretary General Ban Ki-moon launched **SE4ALL** as a global, multi-stakeholder initiative to promote sustainable energy production and consumption to realise the goals of sustainable development, poverty eradication and global prosperity. To achieve these goals, the SE4ALL initiative targets three interconnected objectives by 2030:

- providing universal access to modern energy services
- doubling the global rate of improvement in energy efficiency
- doubling the share of renewable energy in the global energy mix.

SE4ALL identifies the second objective of improving energy efficiency as having the clearest impact on cost reductions, business productivity and increased energy services for all citizens. This acknowledges the fundamental role that energy plays in supporting development by enabling "more from the same" – i.e. achieving greater global resource productivity and greater economic growth from existing energy resources, while reducing the costs to citizens. The UN estimates that energy efficiency savings in developing countries could help broaden energy access while spurring macroeconomic growth. In more industrialised countries, SE4ALL promotes energy efficiency investment for increased energy productivity.

The untapped potential of energy efficiency

Current assessments suggest that, under existing policies, two-thirds of the economically viable[5] energy efficiency potential through 2035, across all sectors, will remain unrealised (Figure 1.3). Many barriers contribute to the limited uptake of energy efficiency opportunities, including information failures, split incentives, subsidised pricing of energy, inadequate pricing of externalities and a shortage of financing.

Because it represents a negative quantity (i.e. energy not expended), energy efficiency is often perceived as an intangible concept. Its value is not always apparent to investors, consumers and policy makers, and its role in enabling achievement of diverse economic and social goals is often obscured.

The actual scope of investment to date – relative to the large potential – indicates that energy efficiency measures are under-valued in the market, by both private sector actors and government policy makers. Even the concept of energy efficiency operating as a market is somewhat novel: experts are only just beginning to understand what such a market might comprise. Improved understanding of the potential for investment in the energy efficiency market is likely to be supported by improved knowledge about the benefits that energy efficiency can deliver in real terms.

5 In the IEA *World Energy Outlook 2012*, investments were classified as "economically viable" if the payback period for the up-front energy efficiency investment is equal to or less than the amount of time an investor might be reasonably willing to wait to recover the cost, using the value of undiscounted fuel savings as a metric. The payback periods used were (in some cases) longer than current averages; but they were always shorter than the technical lifetime of individual assets.

© OECD/IEA, 2014

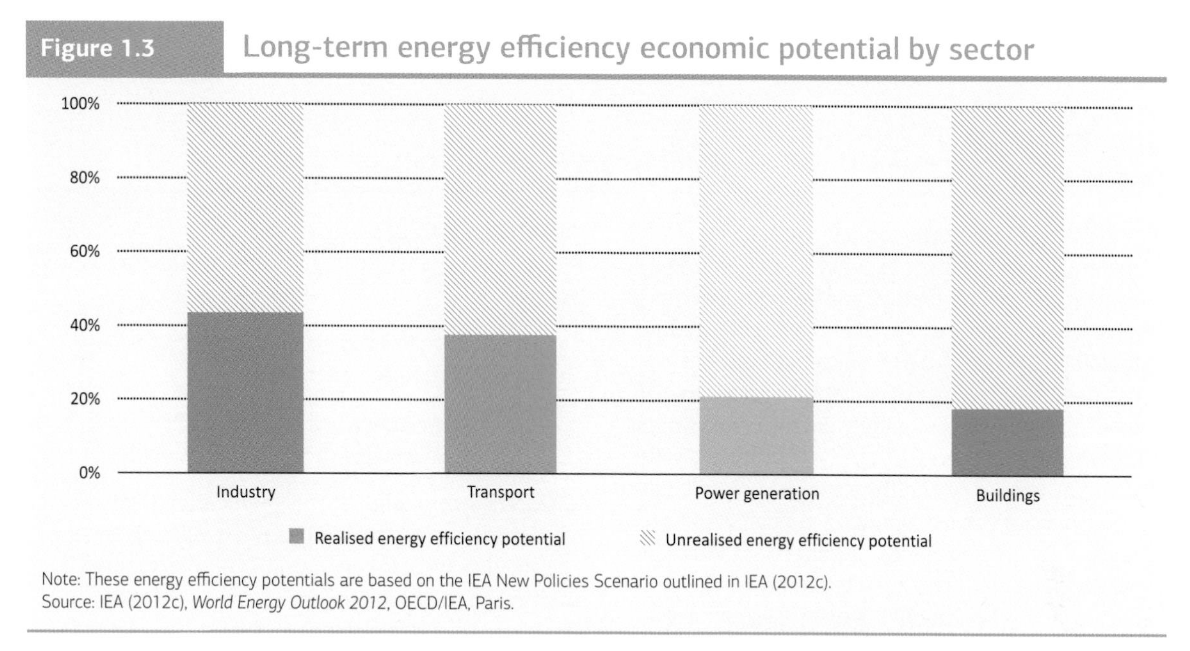

Figure 1.3 Long-term energy efficiency economic potential by sector

Note: These energy efficiency potentials are based on the IEA New Policies Scenario outlined in IEA (2012c).
Source: IEA (2012c), *World Energy Outlook 2012*, OECD/IEA, Paris.

Key point *IEA projections to 2035 show that as much as two-thirds of energy efficiency potential will remain untapped unless policies change.*

A multiple benefits approach adds value to energy efficiency

The multiple benefits approach extends the reach of energy efficiency beyond the well-established impacts of energy demand reduction and reduced GHG emissions, revealing its potential to deliver a host of other benefits to the economy and society. In close consultation with experts from a range of sectors, the IEA has identified numerous benefits that, because of their apparent scale and value to economic and social development, merit further investigation. Thirteen broader benefit areas are listed below, which fall into several thematic categories: energy system security, economic development, social development, environmental sustainability and increasing prosperity.

The brief descriptions below affirm that most are relevant to IEA member countries and non-member countries alike, although prioritisation by individual countries is likely to vary (see Chapter 7). Five of these benefit areas are investigated in depth in the remaining chapters of this publication: macroeconomic development, public budgets, health and well-being, industrial productivity, and energy delivery. These five benefit areas were selected with guidance from IEA member countries (Box 1.3); for each topic, emerging analysis from relevant case studies is highlighted to help demonstrate their value.

Enhancing energy system security

Energy security

Energy efficiency improvements that result in reduced demand can improve the security of energy systems across the four dimensions of risk: fuel availability (geological), accessibility (geopolitical), affordability (economic) and acceptability (environmental and social). While policy makers are alert to the contribution of energy efficiency to improving energy security,

© OECD/IEA, 2014.

the multi-dimensional nature of energy security makes it difficult to quantify; to date, few studies have attempted to examine this link on a comprehensive, economy-wide scale.

Box 1.3 How the IEA selected focus areas in this publication

In 2012, the IEA conducted a broad scoping study to identify the multiple benefits that energy efficiency can deliver both in and beyond the energy sector. This generated a rich menu of potential outcomes linked to energy efficiency policy. It also demonstrated the need to better evaluate the diverse benefits and to widely disseminate results. To stimulate further work on the multiple benefits of energy efficiency, the IEA consulted closely with more than 300 experts working in diverse sectors in which energy efficiency has direct or indirect impacts. The collective agreed to narrow the focus such that this book would draw on best available information from governments and academia to provide in-depth analysis of five benefit areas: macroeconomic development; public budgets; health and well-being; industrial productivity; and energy delivery. These areas were chosen for two reasons:

- they have been identified as policy priorities for IEA member countries
- enough evidence is available about their potential impacts to begin robust analysis.

Chapters 2 through 6 of this publication investigate these benefit areas in turn. This listing is not intended to indicate any ranking of importance within the multiple benefits of energy efficiency since, as discussed elsewhere, different stakeholders will likely prioritise different benefits depending on their country context and other factors.

Energy delivery (covered in Chapter 6)

Improving efficiency within the energy sector can help energy providers deliver better service for their customers while reducing their own operating costs, improving profit margins and mitigating risk. Interesting results have emerged in North America in particular, where regulatory requirements have encouraged in-depth assessment of a range of benefits from energy efficiency programmes. Studies show that NEBs may account for as much as 45% of total programme benefits or even more for programmes targeting low-income households (Synapse Energy Economics, Inc., 2014).

Energy prices

All else being equal, decreased demand for energy services across several markets should prompt a reduction in energy prices (noting that oil prices remain a key driver of energy pricing). The IEA New Policies Scenario projects that world primary energy demand will increase by 40% between 2009 and 2035 (from 12 150 million tonnes of oil-equivalent [Mtoe] to 16 950 Mtoe); by contrast, in the EWS, global demand increases by only 23% over the same period, reaching 14 850 Mtoe. Lower energy prices can influence the economic competitiveness of industry and commerce and reduce the burden of energy costs for individuals, making energy services more affordable and freeing up resources for diverse expenditures (including additional energy consumption).

Economic development

Energy efficiency supports national and international goals for economic development. It enables countries to grow their economies and, in developing countries, to substantially raise standards of living while ensuring that growth is robust and sustainable (Box 1.4).

Macroeconomic development (covered in Chapter 2)

Energy efficiency generally has positive macroeconomic impacts, particularly by boosting GDP and employment. Additional cumulative benefits derive from a range of other impacts,

© OECD/IEA, 2014.

most of which are indirect effects resulting from lower energy expenditures, economy-wide investment in energy efficiency and increased consumer spending. Early evidence suggests a small but positive effect on GDP of 0.25% to 1.1%, depending on the sector and the scale of investment. It has become clear that careful design of modelling approaches is vital to enhance accuracy in this space.

Box 1.4 Energy efficiency for emerging economies: "Doing even more with more"

Improved energy efficiency can play a critical role in supporting development in emerging economies and developing countries. While many industrialised countries prioritise energy efficiency to boost economic output while using fewer energy resources, emerging economies and developing countries seek to accelerate growth by coupling improved efficiency with efforts to expand their resource base – in effect, doing even more with more. Energy efficiency provides a variety of ways for such countries to optimise their resource base to reduce poverty and support sustainable growth.

- **Access**: Energy efficiency can help countries to expand access, effectively enabling them to supply power to more people through existing or expanding energy infrastructure.
- **Development/growth**: Energy efficiency has a variety of positive impacts that support economic growth, for example by improving industrial productivity. One important potential is reducing dependence on imports of oil and other fossil fuels, thereby improving their balance of trade and lowering exposure to associated price and volume volatility.
- **Affordability/poverty alleviation**: Energy efficiency can increase the affordability of energy services (such as lighting, heating, cooling or refrigeration) when it has the effect of reducing the per-unit cost.
- **Local pollution**: Various developing countries are looking to reduce pollutants, notably from local power generation. Energy efficiency (both supply side and end-use) can help to reduce the need for generation – and lower associated pollutants – while supporting economic growth.
- **Climate change mitigation and resilience**: Energy efficiency can enable developing countries to pursue economic growth while limiting national carbon emissions, in particular when coupled with increased renewables and other decarbonisation activities. In addition, by reducing the need for energy infrastructure, energy efficiency reduces the amount of energy assets exposed to extreme weather events, thereby boosting resilience of the energy system as a whole.

The scale of potential multiple benefits of energy efficiency is so large in developing countries that it may overwhelm the technical potential for energy demand reductions. As energy efficiency improves the welfare of these energy users, it is likely to drive a greater rebound effect than is expected in more developed countries (see discussion of the rebound effect in a multiple benefits context below).

Industrial productivity (covered in Chapter 5)

Benefits for industrial firms from energy efficiency measures include enhanced production and capacity utilisation, reduced resource use and pollution, and lower operation and maintenance (O&M) costs, all of which contribute to improved productivity and value creation for the company. Relatively few assessments have attempted to measure these impacts concretely but current results point to additional value ranging from 40% to 250% of the value of energy savings (Lilly and Pearson, 1999; Pearson and Skumatz, 2002).

© OECD/IEA, 2014.

Social development

Poverty alleviation: Energy access and affordability

Recent estimates indicate that 1.2 billion people, primarily in developing countries, are still without access to electricity.[6] As energy suppliers improve their own efficiency, they can provide electricity to more households, thereby supporting increased access to energy. In both developed and developing countries, the poor are more likely to live in inefficient housing and are less able to afford the up-front cost of energy efficiency goods and services; thus, they face higher energy costs than the wealthy. As energy bills are reduced, poorer households have the ability to acquire more and better energy services; they can also spend the freed up income on satisfying other critical needs.

Health and well-being (covered in Chapter 4)

By improving the quality of indoor environments and reducing energy bills, energy efficiency measures in buildings have the potential to significantly improve the health and well-being of occupants. In the residential sector, health and well-being benefits from insulation and heating system improvements have, in many cases, been shown to significantly outweigh energy savings, with benefit-cost ratios as high as 4:1 (Grimes et al., 2011). Health and well-being benefits are consistently strongest among vulnerable groups, including children, the elderly and those with existing illnesses (Box 1.5).

Employment

Investment in energy efficiency can generate a net gain in employment rates both directly and indirectly (through an expenditure shift effect). Reduced unemployment provides a variety of social benefits, in addition to monetary ones, such as improved household incomes and reduced budgetary outlays for unemployment payments. While the value of jobs created can be complicated by various factors such as the labour intensity, geographical location, wage rates and temporal durability, estimates have ranged as high as 27 job years created for every EUR 1 million spent on energy efficiency regulation measures in the residential sector (Wade, Wiltshire and Scrase, 2000). Job creation benefits are investigated in the context of macroeconomic impacts in Chapter 2 of this report.

Box 1.5 A key tool for tackling fuel poverty

Recent estimates suggest more than 150 million people are living in "fuel poverty" in the European Union alone – and that this number is growing (UK ACE, 2014). A household is broadly defined as being in fuel poverty if more than 10% of its annual income is spent on energy. Most often, fuel poverty arises at the nexus of low income, poor housing quality and high energy costs. Fuel poverty is strongly associated with sub-optimal physical and mental health. Energy efficiency retrofits of low-income housing offer a more enduring solution to these problems than energy tariff subsidies or fuel payments because they address the cause of fuel poverty, rather than the symptoms.

Several IEA member countries, including Australia, Ireland, New Zealand, the United States and the United Kingdom have targeted energy efficiency policies to address fuel poverty with positive results. A study using data from New Zealand's *Warm Up NZ: Heat Smart* programme evaluation indicated significantly higher monetised benefits among families on low to modest incomes of USD 519 per year after the retrofitting compared to USD 183 for higher-income families (Telfar et al., 2011).

6 Over 1.2 billion people are without access to electricity and 2.6 billion people lack clean cooking facilities. More than 95% of these people are either in sub-Saharan Africa or developing Asia, and 84% are in rural areas.

© OECD/IEA, 2014.

Environmental sustainability

Local air pollution

Energy efficiency in all sectors can play a major role in reducing outdoor concentrations of local and/or regional air pollutants (such as sulphur dioxide, particulate matter, unburned hydrocarbons and nitrogen oxides); in doing so, it can drive a range of associated economic, environmental and health benefits (see Chapter 4 on the health and well-being impacts of energy efficiency). Measures that support public transport planning, lower-emission motor vehicles and active travel[7] can have particularly strong impacts. In China, for example, the problem of air pollution is particularly acute, due to the rapid increase in the number of motor vehicles and large share of coal-based electricity generation. The total health costs associated with outdoor air pollution in China's urban areas in 2003 was estimated at between USD 25 billion and USD 83 billion (CNY 157 billion and CNY 520 billion),[8] accounting for 1.2% to 3.3% of national GDP (World Bank and SEPA, P. R. China, 2007).

Resource management

At both national and aggregated international levels, lowered energy demand can reduce pressure on scarce natural resources, reducing the need to explore increasingly challenging contexts for extraction (such as ultra-deep offshore, arctic and shale). This can also minimise the related incremental investment costs and environmental uncertainties. Reducing energy consumption and emissions through energy efficiency also plays a role in reducing waste and associated pollution of land and water, thereby contributing to efforts to combat ocean acidification and limit negative impacts on biodiversity.

Increasing prosperity

Public budgets (covered in Chapter 3)

Whether by reducing government expenditures on energy or by generating increased tax revenues through greater economic activity and increased spending on energy efficiency-related and other goods and services, energy efficiency improvements can have important impacts on the budgetary position of national and sub-sovereign entities. Countries that rely on fuel imports are likely to benefit from a related positive impact on currency reserves while domestic energy efficiency in exporting countries can free up additional fuel for export, thereby leading to increased revenue. For countries with energy consumption subsidies, lower consumption reduces government budgetary outlays to finance the subsidies.

Disposable income

Across all income levels, energy efficiency improvements that have the effect of reducing energy bills will increase disposable income for individuals, households and enterprises. The disposable income that could be liberated and spent by consumers is implicitly included in macroeconomic evaluations. How these funds are ultimately spent has an important role in boosting economic activity and also influences the rebound effect.

Asset values

Recent evidence suggests that individuals and businesses are willing to pay a rental and/or sales premium for property with better energy performance. A study of the value of this premium for commercial property indicates that every USD 1 saved in energy costs translates, on average, to acceptance of a 3.5% increase in rent and a 4.9% premium in market valuation (Eichholtz, Kok and Quigley, 2011).

7 Active travel includes any mode of travel that relies on human physical energy as opposed to motorised and carbon-fuelled modes: e.g. walking, running and cycling.

8 CNY = Chinese yuan renminbi.

© OECD/IEA, 2014.

The tangible outcomes of energy efficiency

The list above – which is not exhaustive – gives an overview of the type and scope of potential benefits available through energy efficiency. It demonstrates that energy savings are but one outcome of improved energy efficiency among many of equal or greater value to society.

Communicating the value of energy efficiency is a key challenge. Among non-energy experts, energy is rarely perceived as a commodity (except in terms of the political value of oil), nor it is perceived as a service in the usual sense (e.g. like a mechanic fixing one's car). Energy is the enabler of both goods and services, but remains largely invisible to the general public. In this context, it is useful to discuss both energy and energy efficiency from the perspective of the services they provide. People buy light bulbs not because they want light bulbs but because they want light; they buy washing machines because they want clean clothes. Energy efficiency enhances the value of that service. Shifting the focus from energy efficiency as a good in its own right to what improved efficiency delivers can help stakeholders to grasp its impact and value.

The rebound effect in a multiple benefits context

What is the rebound effect?

One of the most persistent challenges in energy efficiency policy is accounting for the phenomenon known as the "rebound effect" – where improved efficiency is used to access more goods and services rather than to achieve energy demand reduction. As a result, actual energy demand reductions often fall short of the estimates made during the policy development phase.

The rebounding energy consumption poses a problem when energy efficiency policies, which have been implemented on the basis of an expected amount of energy demand reduction, do not deliver the expected results. Policy makers need to fully assess and account for any potential rebound effects when planning energy efficiency policies to ensure that targets are realistic, particularly when other goals that are driven by reduced energy consumption (such as lower GHG emissions) are linked to the policy.

Applying a multiple benefits perspective to energy efficiency deepens understanding of the drivers and impacts of the rebound effect. Increasingly, economists and other analysts demonstrate that energy savings are just one of many outcomes delivered by energy efficiency. A more appropriate weighting of the various outcomes based on national priorities may show that the rebound effect is not always a bad thing; it can often be positive.

What drives the rebound effect?

The rebound effect is generally driven by one of three things:

- the **take-back effect**, where energy users increase their consumption of the energy-using service, rather than accepting the same service at a lower energy or financial cost
- the **spending effect**, where energy users opt to spend their financial savings from reduced energy consumption in the purchase of other energy-consuming activities
- the **investment effect**, where investment in energy efficiency goods and services, stimulated by a policy, lead to an indirect increase in economic activity and energy consumption.

© OECD/IEA, 2014.

An energy user's response to increased insulation in a dwelling is a good example of how the rebound effect occurs. This energy efficiency intervention generates two interdependent outcomes: reduced energy cost and increased indoor temperature. Households will seek to maximise the utility of their energy service by choosing between these two available outcomes based on their personal preferences, needs and behaviours. Budget-constrained households might prefer to take the improved efficiency as a reduction in energy costs (and therefore save energy). Households who struggle to keep a home warm may prefer to take increased comfort (and maintain or increase their energy use). The latter case will drive a rebound effect.

Direct rebound effects can range from 0% (e.g. in whiteware) to as much as 65% (e.g. electrically heated homes in California) (Hertwich, 2005). However, estimates tend to converge between 10% and 30% (Sorrell, 2007; Jenkins, Nordhaus and Shellenberger, 2011).

The implication of the rebound effect for multiple benefits

A multiple benefits approach to energy efficiency makes it possible to consider what a rebound effect represents in terms of welfare improvements for society. All impacts of the rebound effect must be understood against the backdrop of the particular energy efficiency policy and the economic context as well as priority outcomes, so that policies can be designed to achieve prioritised objectives and address undesirable elements of the rebound effect (Rebound effect perspective 1).

Some benefits come with an energy consumption price tag (e.g. when improved energy affordability leads to increased energy consumption for heating). Where energy savings are taken back in the achievement of health benefits, poverty alleviation, improving productivity or reducing supply-side losses, the rebound effect created can be viewed as a net positive outcome, amplifying the benefits of the energy efficiency intervention. Often a rebound effect actually signals a positive outcome from the perspective of broader economic and social goals. In some contexts, such as developing countries with high growth rates where activity tends to be more energy-intensive, rebound may often be desirable as it enables the economy to capitalise further on its energy resources and stimulate other efficiencies.

One must consider the rebound effect against the backdrop of the specific benefit(s) being targeted. Unbundling the relationship between energy savings and the broader outcomes of energy efficiency can provide a fuller understanding of the rebound effect, and a clearer appreciation of whether this effect reduces or amplifies the benefits of an energy efficiency intervention.

Empowering policy makers to assess multiple benefits

A multiple benefits approach can provide a deeper understanding of the dynamics at play in any investment in and improvement of energy efficiency. Once these are understood, policy design and implementation can be adjusted to minimise undesirable impacts and maximise prioritised impacts. Ultimately, it is the task of policy makers to consider any trade-offs between implementation and other costs and socio-economic welfare gains of a particular energy efficiency policy.

To effectively inform policy decisions, multiple benefits need to be supported by a robust evidence base, which implies a rigorous approach to gathering data, quantifying the benefits and applying study results to address policy challenges. A multiple benefits approach has implications for all phases of the policy process, from planning to implementation, monitoring and evaluation. The Companion Guide at the end of this publication presents methodologies and tools that can be used to implement a multiple benefits approach within a national policy process.

© OECD/IEA, 2014.

Rebound effect perspective 1

Understanding the dynamics of the rebound effect: When does it matter?

To estimate the benefits that arise as a result of an energy efficiency policy, it is essential to accurately calculate the energy saved by taking into account the rebound effect. To properly understand a given rebound effect, it is necessary to take into account the dynamics created by energy efficiency policy across the economy.

There are two aspects to this issue. One is the role of different benefits in driving the rebound effect: i.e. some multiple benefits may come with an energy consumption price tag and therefore generate a rebound effect; others, such as environmental impacts, have no marked impact on energy consumption. The other aspect is the impact of the rebound effect on achieving multiple benefits, i.e. some multiple benefits are a flow-on effect from energy savings, in which case the rebound effect will have implications for achieving those benefits. Other benefits occur independently of the rebound effect.

It is helpful to identify the multiple benefits that rely on energy savings in order to ascertain which will be affected by a rebound effect. While many of the multiple benefits presented are linked to energy savings to some extent and any rebound effect will impinge on their magnitude, others are largely unaffected (Table 1.1). The second row of Table 1.1 indicates whether different multiple benefits can raise energy consumption and have the potential to cause a rebound effect. The multiple benefits described in this book fall differently across the categories outlined in the table below.

Table 1.1 The interaction between multiple benefits and energy savings

Benefit	Energy security	Energy delivery	Energy prices	Macroeconomic development	Industrial productivity	Poverty alleviation	Health and well-being	Employment	Environmental sustainability	Public budgets	Disposable income	Asset values	Development
Entirely dependent on energy savings	Y	Y	Y	N	N	Y	N	N	Y	N	Y	N	N
Could drive up energy consumption	N	N	Y	Y	Y	Y	Y	Y	N	Y	Y	N	Y

Track these rebound effect perspectives in subsequent chapters for insights into how the rebound effect relates to the macroeconomic, public budget, health and well-being, industrial productivity and energy delivery benefits of energy efficiency.

Some cautions for the multiple benefits approach

Multiple benefits are dynamic and present several complexities. Experience to date has identified several issues that are important to consider when applying a multiple benefits approach to assessment of energy efficiency policies.

- The impacts of energy efficiency can be both positive and negative; in some cases, some benefits may come with additional costs. A holistic and comprehensive analytical approach is needed to capture both positive and negative impacts in each case, and enable evaluators to assess whether the net effect of a particular energy efficiency policy is positive or negative.
- Categorising the benefits can help to better understand their characteristics and scope. Benefits in the industrial sector, for example, have been categorised into competitiveness, production, O&M, working environment, and environmental compliance. Benefits from the perspective of energy utilities can be separated into direct and indirect, but have also been

© OECD/IEA, 2014.

further divided in some cases. For example, in the United States cost-effectiveness tests apply an economy-wide perspective to distinguish between participant (customer), utility and societal benefits. The appropriate categorisation is likely to vary depending on the perspective being applied or the context in which the multiple benefits are being measured, but will also have an impact on how the benefits are understood and evaluated.

- Some benefits are indirect, or are the product of a chain of actions. Detailed investigation may be required to identify which action stimulates a specific benefit. Decomposition analysis, for example, could be used to attribute a specific benefit to a particular energy efficiency measure.
- As noted above, energy efficiency measures can impact all levels of the economy – from individual citizens or households to sectoral or sub-sectoral – at the national or international scale. Impacts can even flow across levels or across borders to achieve shared goals. A corollary of this is that a direct impact in one area of the economy can have flow-on effects in another; as a result, impacts can arise in different areas at the same time (or at different times). It is therefore extremely important to establish mechanisms to avoid double-counting these impacts when carrying out final benefit-cost analysis of specific measures or of programmes as a whole.

Conclusion

Evidence currently available indicates that energy efficiency can have a range of potentially large and valuable impacts across many areas of economic and social development. At present, most technical assessments and political discussions about energy efficiency maintain a narrow focus on traditional measures of reduced energy demand and fewer GHG emissions. As a result, most energy efficiency policies fall short of capturing the true potential of this energy resource (the "hidden fuel") that, despite being on scale with fossil fuel sources and preferable from a sustainability perspective, remains largely untapped.

The proposed multiple benefits approach can help to address this shortcoming. Raising awareness of the specific energy efficiency benefits across diverse sectors is a first step to support the more complete consideration of energy efficiency as a mainstream tool for both economic and social development. There is, of course, need for caution in moving into this emerging field. Stakeholder involvement is needed to contribute to building knowledge and strengthening experience.

Presenting the mechanisms by which a broader range of benefits can be measured and monetised, and demonstrating how they can be integrated into policy development and evaluation, will support policy makers in their efforts to optimise the potential value of energy efficiency.

Bibliography

ACE (Association for the Conservation of Energy) (2014), *The Energy Bill Revolution. Fuel Poverty: 2014 Update*, ACE, London, www.e3g.org/docs/ACE_and_EBR_fact_file_%282014-02%29_Fuel_Poverty_update_2014.pdf (accessed 13 May 2014).

Anderson, D.R. et al. and Health Enhancement Research Organization (HERO) Research Committee (2000), "The relationship between modifiable health risks and group level health care expenditures", *American Journal of Health Promotion*, Vol. 15, No. 1, AJPH (American Journal of Health Promotion) and Allen Press, North Hollywood, pp. 45-52, www.ncbi.nlm.nih.gov/pubmed/11184118 (accessed 24 June 2014).

APERC (Asia Pacific Research Centre) (2007), *A Quest for Energy Security in the 21st Century: Resources and Constraints*, APERC, Tokyo, http://aperc.ieej.or.jp/file/2010/9/26/APERC_2007_A_Quest_for_Energy_Security.pdf (accessed 23 June 2014).

© OECD/IEA, 2014.

Eichholtz, P., N. Kok and J.M. Quigley (2011), *The Economics of Green Building*, Program on Housing and Urban Policy Working Paper Series No. W10-003, University of California, Berkeley, Berkeley, http://urbanpolicy.berkeley.edu/pdf/EKQ_041511_to_REStat_wcover.pdf (accessed 29 June 2014).

EC (European Commission) (2012), "Directive 2012/27/EU of the European Parliament and of the Council of 25 October 2012 on energy efficiency, amending Directives 2009/125/EC and 2010/30/EU and repealing Directives 2004/8/EC and 2006/32/EC Text with EEA relevance", *Official Journal of the European Union*, L series, Publications Office of the European Union, EUR-Lex, pp. 1-56, http://eur-lex.europa.eu/legal-content/EN/ALL/?uri=OJ:L:2012:315:TOC (accessed 29 June 2014).

Grimes A. et al. (2011), *Cost Benefit Analysis of the Warm Up New Zealand: Heat Smart Program*, Report for Ministry of Economic Development, Motu Economic and Public Policy Research, Wellington (revised June 2012).

Hanley, N. et al. (2009) "Do increases in energy efficiency improve environmental quality and sustainability?", *Ecological Economics*, Vol. 68, No. 3, Elsevier Ltd., Amsterdam, pp. 692-709, www.sciencedirect.com/science/article/pii/S0921800908002589 (accessed 29 June 2014).

Hayes, S., R. Young and M. Sciortino (2012), *The ACEEE 2012 International Energy Efficiency Scorecard*, ACEEE (American Council for an Energy Efficient Economy) Report No. E12A, Washington DC.

Hertwich, E.G. (2005), "Consumption and the rebound effect: an industrial ecology perspective", *Journal of Industrial Ecology*, Vol. 9, No. 1-2, Wiley-Blackwell Publishing, Hoboken, pp. 85-98, http://onlinelibrary.wiley.com/doi/10.1162/1088198054084635/pdf (accessed 29 June 2014).

Hills, J. (2012), *Getting the Measure of Fuel Poverty: Final Report of the Fuel Poverty Review*, CASE *Report* 72, commissioned by UK DECC (United Kingdom, Department of Energy and Climate Change), Crown copyright, CASE (Centre for the Analysis of Social Exclusion), London, http://sticerd.lse.ac.uk/dps/case/cr/CASEreport72.pdf (accessed 28 June 2014).

IEA (International Energy Agency) (2014), *Energy Technology Perspectives 2014 – Harnessing Electricity's Potential*, OECD (Organisation for Economic Co-operation and Development)/IEA, Paris, www.iea.org/etp/etp2014 (accessed 29 June 2014).

IEA (2013), *Energy Efficiency Market Report 2013: Market Trends and Medium-Term Prospects*, OECD/IEA, Paris, www.iea.org/w/bookshop/add.aspx?id=460 (accessed 19 June 2014).

IEA (2012a), *Spreading the Net: The Multiple Benefits of Energy Efficiency Improvements*, Insights Paper, OECD/IEA, Paris, www.iea.org/publications/insights/insightpublications/name,26319,en.html (accessed 18 June 2013).

IEA (2012b), *Plugging the Energy Efficiency Gap with Climate Finance: The Role of International Financial Institutions (IFIs) and the Green Climate Fund to Realise the Potential of Energy Efficiency in Developing Countries*, Insights Paper, OECD/IEA, Paris, www.iea.org/publications/freepublications/publication/PluggingEnergyEfficiencyGapwithClimateFinance_WEB.pdf (accessed 19 June 2014).

IEA (2012c), *World Energy Outlook 2012*, OECD/IEA, Paris, www.iea.org/publications/freepublications/publication/name-49561-en.html (accessed 18 June 2014).

IEA (2011a), *25 Energy Efficiency Policy Recommendations: 2011 Update*, OECD/IEA, Paris, www.iea.org/publications/freepublications/publication/name,3782,en.html (accessed 18 June 2018).

IEA (2011b), *World Energy Outlook 2011*, OECD/IEA, Paris, www.iea.org/publications/freepublications/publication/name,37085,en.html (accessed 18 June 2014).

IEA (2011c), *Saving Electricity in a Hurry*, Energy Efficiency Series, OECD/IEA, Paris, www.iea.org/publications/freepublications/publication/Saving_Electricity.pdf (accessed 23 June 2014).

IEA (2010), *Energy Efficiency Governance*, OECD/IEA, Paris, www.iea.org/publications/freepublications/publication/name-3936-en.html (accessed 18 June 2014).

Jenkins, J., T. Nordhaus and M. Schellenberger (2011), *Energy Emergence: Rebound and Backfire as Emergent Phenomena*, Breakthrough Institute, Oakland, http://thebreakthrough.org/archive/new_report_how_efficiency_can (accessed 23 June 2014).

Kruyt, B. et al. (2009), "Indicators for energy security", *Energy Policy*, Vol. 37, No. 6, Elsevier Ltd., Amsterdam, pp. 2166-2181, www.sciencedirect.com/science/article/pii/S0301421509000883 (accessed 23 June 2014).

© OECD/IEA, 2014.

Lilly, P. and D. Pearson (1999), "Determining the full value of industrial efficiency programs", *Proceedings from the 1999 ACEEE Summer Study on Energy Efficiency in Industry*, Saratoga Springs, June 1999, ACEEE (American Council for an Energy Efficient Economy), Washington DC, pp. 349-362, www.seattle.gov/light/Conserve/Reports/paper_7.pdf (accessed 22 June 2014).

Matthews, C. et al.(2006), "Influence of exercise, walking, cycling, and overall nonexercise physical activity on mortality in Chinese women", *American Journal of Epidemiology*, Vol. 165, No. 12, Johns Hopkins Bloomberg School of Public Health, Baltimore, pp. 1343-1350, http://aje.oxfordjournals.org/content/165/12/1343.full.pdf (accessed 28 June 2014).

Pearson, D. and L. Skumatz (2002), "Non-energy benefits including productivity, liability, tenant satisfaction, and others – what participant surveys tell us about designing and marketing commercial programs", *Proceedings of the 2002 Summer Study on Energy Efficiency in Buildings*, Vol. 4, ACEEE (American Council for an Energy Efficient Economy), Washington DC, pp. 4.289-4.302 www.aceee.org/files/proceedings/2002/data/index.htm (accessed 23 June 2014).

Rabl A. and A. Nazelle (2011), "Benefits of shift from car to active transport", *Transport Policy*, Vol. 19, No. 1, Elsevier Ltd., Amsterdam, pp. 121-131, www.sciencedirect.com/science/article/pii/S0967070X11001119 (accessed 28 June 2014).

Skumatz, L. (1997), "Recognizing All Program Benefits: Estimating the Non-Energy Benefits of PG&E's Venture Partner's Pilot Program (VPP)", *Proceedings of the 1997 Energy Evaluation Conference*, Chicago, Illinois, 1997.

Sorrell, S. et al. (2004), *The Economics of Energy Efficiency: Barriers to Cost-Effective Investment*, Edward Edgar Publishing, Cheltenham, www.e-elgar.co.uk/bookentry_main.lasso?id=2607 (accessed 18 June 2014).

Sorrell, S. (ed.) (2007), *The Rebound Effect: an Assessment of the Evidence for Economy-wide Energy Savings from Improved Energy Efficiency*, report by the Sussex Energy Group for the Technology and Policy Assessment function of the UKERC (UK Energy Research Centre), London, www.ukerc.ac.uk/support/ReboundEffect (accessed 18 June 2014).

Strauss S., S. Rupp and T. Love (eds.) (2013), *Cultures of Energy: Power, Practices, Technologies*, Left Coast Press, Walnut Creek, www.lcoastpress.com/book.php?id=407 (accessed 18 June 2014).

Synapse Energy Economics, Inc. (2014), Unpublished material provided to the IEA, 3 March 2014.

Taylor, R.P. et al. (2008), *Financing Energy Efficiency: Lessons from Brazil, China, India and Beyond*, The World Bank, Washington DC, https://openknowledge.worldbank.org/handle/10986/6349 (accessed 18 June 2014).

Telfar-Barnard, L. et al. (2011), *The Impact of Retrofitted Insulation and New Heaters on Health Services Utilisation and Costs, Pharmaceutical Costs and Mortality: Evaluation of Warm Up New Zealand: Heat Smart*, report to the MED (Ministry of Economic Development), MED, Wellington, www.motu.org.nz/ (accessed 21 June 2014).

Turner, K. (2013), "'Rebound' Effects from Increased Energy Efficiency: A Time to Pause and Reflect", *The Energy Journal*, Vol. 34, No. 4, IAEE (International Association for Energy Economics), Cleveland, www.iaee.org/en/publications/ejarticle.aspx?id=2523 (accessed 18 June 2014).

UNFCCC (United Nations Framework Convention on Climate Change) Secretariat (2014), "Work plan on enhancing mitigation ambition technical expert meeting to unlock mitigation opportunities in energy efficiency in pre-2020 period", summary at the closing session of the technical expert meetings by the facilitator Mr. Jun Arima (Japan), Bonn, 10-12 March 2014, http://unfccc.int/files/bodies/awg/application/pdf/tem_on_ee_summary.pdf (accessed 3 May 2014).

Wade, J., V. Wiltshire and I. Scrase (2000), *National and Local Employment Impacts of Energy Efficiency Investment Programmes: Final Report to the Commission April 2000 Volume 1: Summary Report*, Association for the Conservation of Energy, London, www.ukace.org/wp-content/uploads/2012/11/ACE-Research-2000-04-National-and-Local-Employment-Impacts-of-Energy-Efficiency-Investment-Programmes-Volume-1-Summary-Report.pdf (accessed 18 June 2014).

World Bank and SEPA (State Environmental Protection Administration), P. R. China (2007), *Cost of Pollution in China: Economic Estimates of Physical Damages*, World Bank, Washington DC, http://siteresources.worldbank.org/INTEAPREGTOPENVIRONMENT/Resources/China_Cost_of_Pollution.pdf (accessed 28 June 2014).

WRI (World Resources Institute) et al. (2014), *Energy Efficiency in U.S. Manufacturing: The Case of Midwest Pulp and Paper Mills*, WRI, Washington DC, http://pdf.wri.org/energy-efficiency-in-us-manufacturing-midwest-pulp-and-paper.pdf (accessed 18 June 2014).

© OECD/IEA, 2014.

Chapter 2

Macroeconomic impacts of energy efficiency

Key points

- *The analysis of benefits in this publication begins with macroeconomic impacts because it is an overarching impact and priority in many situations. This area has been extensively analysed by many policy makers and economists and the state of analysis is relatively mature, but its systematic integration into energy efficiency policy assessment requires strengthening.*
- *Valuing the full range of multiple benefits of energy efficiency at the macroeconomic level challenges the conventional relationship between energy performance and economic growth: where previously economic performance drove energy consumption upwards, reduced energy consumption now appears to have substantial positive impacts for economic development.*
- *Through investment and cost reduction effects, the macroeconomic benefits of energy efficiency measures manifest most prominently in the areas of increased economic activity (as measured by gross domestic product [GDP]), higher employment, advantageous price impacts and favourable trade balances.*
- *In terms of economic activity, changes due to energy efficiency show GDP growth rates ranging from 0.25% to 1.1%. On the employment front, one modelling study shows that each EUR 1 million invested in energy efficiency creates 8 to 27 job years. The situation for trade balances is less clear; likely energy efficiency will create winners and losers within energy trade.*
- *The complexity of macroeconomic assessment is best met by combining bottom-up (technical) and top-down (economic) models; in many cases, several models may be needed or evaluators may need to develop additional purpose-built models. In the meantime, estimates can suffice for initial assessments.*
- *Macroeconomic assessment raises an interesting question as to whether the future economy may be transformed such that it is driven by cost-saving efficiency gains, rather than the current growth model of consumer spending to exploit cheap energy.*

Introduction: Increasing evidence of macroeconomic benefits

Discussion of multiple[1] benefits in specific areas begins with macroeconomic impacts because they constitute an overarching aspect of much policy analysis and are the priority in many situations. An already significant and growing body of research shows that energy demand reduction from energy efficiency improvements can deliver wider impacts[2] across

1 In other literature, these impacts have been variously labelled "co-benefits", "ancillary benefits" and "non-energy benefits" – and are often used interchangeably with "multiple benefits". The International Energy Agency (IEA) uses the term multiple benefits, which is broad enough to reflect the heterogeneous nature of outcomes and to avoid pre-emptive prioritisation of various benefits; different benefits will be of interest to different stakeholders.

2 An impact is any kind of result from an action or measure, whether positive or negative. It is distinct from a benefit, which typically refers to an impact that is positive. See Annex A: Glossary for more information.

© OECD/IEA, 2014.

the whole economy – i.e. the macroeconomic benefits of energy efficiency. This would appear to turn the conventional relationship between energy performance and economic growth on its head: where previously economic performance drove energy consumption upwards, now measures to reduce energy consumption through energy efficiency improvements appear to have positive impacts for economic development.

Major macroeconomic issues relate to the extent to which energy efficiency measures can reduce energy costs, thereby increasing the competitiveness of the economy and resulting in improved trade balances for energy-consuming countries. Energy efficiency improvements could even drive a shift from demand-driven economies to a new supply-driven model in which efficiency (rather than consumption) is the driver of growth.

Politicians and the general public show a keen interest in economic development and growth issues, especially in the current climate of economic recession and relatively high energy prices in many countries. In fact, policy makers are under pressure to estimate the potential impact of energy efficiency improvements on the wider economy, and are expected to carry out *ex ante* (pre-implementation) assessment or appraisal of a variety of energy efficiency policies. To present credible results, any macroeconomic assessment must be carried out as rigorously and transparently as possible. It should also be supported by sound analysis of public and private benefits and costs at the microeconomic level.

Macroeconomic assessment is a mainstream branch of economic analysis that has built up a huge body of knowledge and evidence over many years. However, the way in which energy efficiency policies and their impacts influence macroeconomic performance still needs to be better understood by many policy makers.

In investigating the impacts of energy efficiency on the wider economy, this chapter aims to provide guidance on how to better assess these impacts. While avoiding complex and more theoretical arguments, the chapter provides some guidance on different estimation methods and the situations in which their application is most suitable.

After outlining a framework to describe the macroeconomic impacts of energy efficiency in relation to both investment and cost reduction effects, the text explores results in assessing impacts on economic output, employment, trade balance and energy price changes. The subsequent section highlights both simple and complex assessment methods (with examples) and then shows a step-by-step process for modelling macroeconomic impacts. A penultimate section addresses areas of concern to policy making, particularly the effects of energy price, crowding out and rebound.

Framing the macroeconomic impacts

The macroeconomy, by definition, covers a broad range of sectors. In this chapter, the term macroeconomic is used to cover economy-wide effects that occur at national, regional and international level in relation to impacts that result from energy efficiency policies. This covers the aggregate effects of energy efficiency measures that comprise both the sum of individual effects and the impacts resulting from complex interactions throughout the economy.

In general, the macroeconomic impacts of energy efficiency are the product of two types of effects associated with energy efficiency measures:

- **investment effects** being the results derived from increased investment in energy efficiency goods and services

© OECD/IEA, 2014.

- **energy demand or cost reduction effects**, which comprise the effects arising from the energy demand reduction (or reduced costs) associated with actually realising an improvement in energy efficiency.

In many cases, investing in energy efficiency goods and/or services is the first action taken as part of an energy efficiency measure,[3] which may lead to the second step of actually realising energy savings from the new good or service. These two steps separately bring about an array of direct and indirect effects that collectively cause macroeconomic impacts (Figure 2.1). Further examination of these categories in the context of macroeconomic impacts will serve to clarify the distinctions and overlaps.

Figure 2.1 Energy demand reduction effects

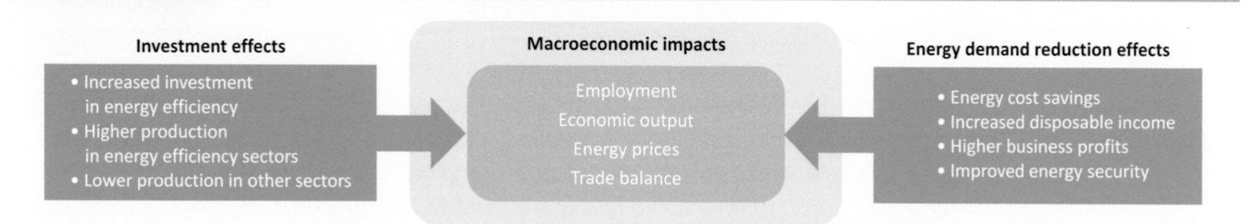

Source: Unless otherwise noted, all material in figures and tables in this chapter derives from IEA data and analysis.

Key point *Macroeconomic impacts are driven by two kinds of effects: investment and energy demand reduction.*

An important consideration is that the economic effects of energy efficiency measures are different for final consumers (i.e. households) and energy-using producers (i.e. businesses). For final consumers, increased energy efficiency can lead to a demand shift from energy consumption to other goods. The producing sectors (business consumers) are more likely to see a benefit in more competitive production (Turner, 2013).

For the final consumers (households), it is relatively straightforward to understand how investment in energy efficiency and the resulting energy demand reduction can lead to increased spending and economic activity. Somewhat less obvious, but still significant, are the ways in which this can lead to indirect (or second-round) effects such as employment, government revenue and price effects (if other investment and spending is not crowded out). Households are likely to experience positive income effects, unless household wage demands increase as the labour supply becomes more competitive.

Categorising the energy efficiency impacts of the producing sectors is more complicated. As businesses can act both as producers and consumers, they can experience a range of effects. The direct effects could include: their own investment in energy efficiency; a rise in demand for energy efficiency goods or services; and increased productivity as a result of reduced energy costs. Second-round effects in the wider economy may also have impacts, including: increased employment in sectors delivering energy efficiency-related goods and services; reduced employment in energy-intensive industries; and multiplier

3 Energy efficiency improvements can also result without investment in technologies or services designed to enhance energy efficiency. It may arise, for example, through behavioural change, in which case only the energy cost reduction effects apply. The large-scale improvement needed to optimise energy efficiency potential globally will require both investment and behavioural change. Thus, the investment effect will apply for most governments seeking to estimate the macroeconomic effects of energy efficiency measures.

© OECD/IEA, 2014.

effects.[4] Understanding the relationship between effects and impacts is a good starting point.

Energy efficiency investment effects

The recently published *IEA Energy Efficiency Market Report* estimates that total global investment in energy efficiency measures in 2011 reached USD 300 billion and the likelihood is that **investment in energy efficiency**[5] will continue to rise as more countries implement energy efficiency policies. Macroeconomic effects from increased investment in energy efficiency arise mainly from the new or reallocation of spending to sectors producing energy efficiency goods and services. The direct effect of this is **higher production levels** leading to increased ***economic output*** and ***employment*** in sectors delivering energy efficiency-enhancing services and products.

For the consumer, household or business investing in energy efficiency, there is an initial up-front outlay (unless policy measures such as grants or pay-as-you-save schemes are in place) that will affect their ability to make other investments and/or undertake consumption in the short run. For businesses, this may lead to stranded assets.[6] Therefore, investment in energy efficiency measures is likely to be offset, to some extent, by **less investment in other sectors**, with associated production, employment and second- and third-round effects. For example, there may be less investment in energy-intensive industries, leading to job losses in those sectors.

Over time, the investment effect may directly influence the ***trade balance***. If the increase in demand for energy efficiency goods and services leads to economies of scale, substituting energy demand for improved energy efficiency and services, this can lead to an improvement in competitiveness and technology innovation in the sectors involved. Energy efficiency measures could also cause negative trade effects if they increase costs for firms and decrease **firm competitiveness**. However if the investment is cost-effective, it should not reduce the competitiveness of a business.

There may be indirect (i.e. second- and third-round) **re-spending effects** that lead to an increase in overall ***economic output***. This could arise if supply chains increase production and employment rises in energy efficiency-related and other sectors, causing demand-based growth as newly employed individuals spend their wages across the economy.

Energy demand reduction effects

When energy efficiency improvements reduce energy demand and costs, the macroeconomic effect of **energy demand reductions** (or cost savings) can influence the entire economy.[7] The **increased disposable income** (for individuals or households) or **higher profits** (for businesses) are then available to be spent on other goods (by consumers; the **spending effect**), or in the case of the producing sectors can be reinvested in the business (**reinvestment effect**) or passed through as lower output prices to consumers (**price effects**), potentially driving growth in ***economic output***.

Estimations of the spending effect show it is substantially larger than the investment effect. It has been estimated that the spending effect accounts for as much as 88% to 91%

4 A multiplier effect indicates when any new injection of spending into the economy, with implications for increased spending (a rise in incomes, etc.) results in an increase in final income. This is related to, but distinct from, the shortcut tool known as a multiplier, which simulates this effect in estimations. See Annex A: Glossary for more information.

5 Effects are in bold in the text and macroeconomic impacts are bold italics so the reader can distinguish between the two.

6 Stranded assets are assets that no longer are in use but are recorded on the balance sheet as a loss in profit. Stranded assets arise when, for example, a power plant or factory is no longer financially viable because new environmental regulations require a higher standard of environmental performance.

7 This assumes that energy prices do not increase.

© OECD/IEA, 2014.

(depending on the fuel) of the ***macroeconomic impacts*** (in terms of gross product for the provinces examined in Canada) that occur over the lifetime of the energy efficiency measures (Howland et al., 2009). These findings were corroborated by a second macroeconomic modelling study of the Eastern Canadian provinces, which found that in a scenario where energy efficiency measures reduced energy consumption by 13% to 23% (depending on fuel type), it would increase GDP by USD 78 713 billion and boost employment by 625 110 job years over the period 2012-40 (compared to a BAU scenario) (ENE, 2012). Further analysis showed that 70% to 90% of the overall GDP and employment impacts could be attributed to the energy cost reduction effects in this region of Canada.

Energy efficiency measures in the producing sectors may lead to **reduced costs of production**, which could lead to output price falls and give a ripple effect of falling costs throughout the economy, thereby creating productivity-led growth. This drives new production and purchase of inputs, some of which will come from local regions. ***Employment*** can be generated when the spending, reinvestment and price effects lead to increased economic activity in energy efficiency-related and other sectors.

Reduced ***energy prices*** can also result from reduced energy consumption and costs, if the energy efficiency activities are sufficiently widespread and of a large enough scale. Some energy sources (such as oil) are global commodities; change in demand in only one region may not have a significant impact on energy prices. Local supply constraints may, however, translate into changes in energy prices locally if energy efficiency measures free the supply of the energy sources and lead to improved **security of energy supply** (Box 2.1). Countries can maintain positive economic growth while decoupling energy use, as energy efficiency can help mitigate higher energy prices.

Box 2.1 Energy security

Increased energy security is a commonly cited economy-wide benefit of energy conservation and efficiency, and warrants mention in the context of macroeconomic benefits. Energy security is defined by the IEA as "the uninterrupted physical availability of energy at a price which is affordable, while respecting environmental concerns." The IEA model of short-term security and those presented by others (Cherp and Jewell, 2011; Scheepers et al., 2007) consider three aspects of energy security: ***robustness*** (adequacy and reliability of resources and infrastructure); ***sovereignty*** (the degree of exposure to threats from foreign actors); and ***resilience*** (ability to respond to diverse disruptions). Energy efficiency models tend to calculate energy imports as an output and therefore focus on sovereignty and resilience.

At present, over 70% of Germany's primary energy demand (PED) is met through energy imports, which make up 11% of German imports overall. Energy imports reached EUR 102 billion in 2012 (Statistisches Bundesamt, 2013). An ambitious energy efficiency strategy to reduce primary energy demand by 6% in 2020 could reduce energy imports by EUR 4.3 billion (Lehr, Lutz and Edler, 2012). The authors of this study stress that energy efficiency measures would do more than simply improve the trade balance; they would also provide security against disruptions, as many of the countries supplying fossil fuels have a tendency to conflict (such as the recurrent disruptions between Russia and the Ukraine). The authors reiterate that energy efficiency is the surest source of energy supply that exists.

Many studies modelling energy efficiency measures include energy security as a key rationale for investing in energy efficiency. However, there are few examples where more analysis is carried out. It is simply assumed that any reduction in energy demand will improve a country's energy security. While this is true generally, analysis needs to be undertaken in each case to understand which energy fuels are likely to be saved through energy efficiency measures, as energy security is likely to improve mainly when non-domestic energy sources are affected.

© OECD/IEA, 2014.

Electricity prices can be reduced as each kilowatt hour saved can reduce the volume of fuel used, greenhouse gas emissions and investment costs for fossil and renewable power plants, and for power grid expansion (see Chapter 6). A study of electricity markets in Germany demonstrates the potential scale of this effect: a 10% to 35% reduction in electricity consumption by 2035 (compared to BAU) would lower electricity generation costs by USD 13.7 billion to USD 27.3 billion (EUR 10 billion to EUR 20 billion) (Wünsch et al., 2014).

If energy prices do fall, further positive impacts on competitiveness are likely for the economy, leading to increased ***economic output*** and greater **disposable income** for consumers.[8] In time, global energy markets may be more closely linked, at which point if energy efficiency measures reduce energy commodity prices, all countries would gain.

Macroeconomic impacts

Four main macroeconomic impacts are typically estimated:

- economic development, measured by GDP
- employment
- energy price changes
- trade balance.

Understanding these four impacts, including what they cover, the metrics needed to measure them and specific considerations is vital to carrying out robust assessments (Table 2.1). The variables and their assessment are elaborated further in the next section, along with identification of useful indicators and some sample assessments from existing literature.

How and whether energy efficiency measures influence these impacts and deliver long-term positive outcomes depends on the structure and nature of the economy, the scale and substance of realised energy efficiency effects, and the distribution of impacts across consumers and producers.

The level of economic development has a strong influence on demand for energy services. Countries with less economic development are likely to have a pent-up demand for energy services. Energy efficiency improvements increase the availability of energy services and therefore can cause macroeconomic impacts disproportionately, depending on the initial level of development (Toman and Jemelkova, 2003). Examining how energy efficiency improvements affected the British economy during the First and Second Industrial Revolutions provides insights for the potential implications of energy efficiency improvements and price reductions of energy service on economic growth. The British economy benefited from a series of declines in energy service prices (focusing on iron smelting, industrial power, and land and sea freight) mostly driven by improvements in energy efficiency. Their influence on growth varied considerably at different levels of economic development (Fouquet, 2014b).

Lower energy prices and major improvements in energy efficiency appear to have had (and are likely to have) major influences on economic growth and development (including

8 Final energy demand reductions may not always lead to price reductions. In a closed electricity market facing exogenously determined input fuel costs, electricity prices may even rise a little, owing to the need to spread the fixed costs of the electricity system over fewer units sold. Whether this effect is actually positive on retail prices will depend on the extent to which marginal plants are more costly, and the time of day of demand reductions (i.e. peak versus off-peak).

© OECD/IEA, 2014.

possibly changing the nature of production and consumption processes). In addition, the energy services that kick-start major periods of economic growth change over time and with levels of economic development (Box 2.2).

Table 2.1 Overview of macroeconomic indicators for energy efficiency impact assessment

Impacted area	Impact	Description	Metrics required	Comments
Economic output (increase)	GDP	The total market value of all final goods and services produced in a country in a given year.	■ Consumer spending ■ Investment ■ Exports – imports ■ Government spending	Impact on GDP is usually modelled at the national or regional level, allowing characterisation of the interactions of energy production, labour markets, economic structure and historical energy efficiency policies. Since GDP comprises an aggregate of many variables across the economy, the impact of energy efficiency activities is likely to be measurable only if the investment or energy efficiency improvement is large and has significant multiplier effects.
Employment (increase)	Number of net new jobs	The number of net jobs that are created or lost, directly or indirectly, through energy efficiency measures. Also the structural shifts in employment that may occur because of energy efficiency measures.	■ Net new jobs ■ Sectoral job shifts ■ Wage rates ■ Labour intensity ■ Local content	One of the key impacts for policy makers and politicians. Estimation should be as transparent and robust as possible and should cover net employment, taking into account both gains and losses. Good sectoral analysis is needed to properly understand current spare labour capacity, skills available, and changes in labour rates for relevant sectors.
Energy price (decrease)	Energy unit price	The consequences are energy cost reduction for net importing countries and consumers and businesses.	■ Cost per unit of energy ■ Energy substitution options ■ Market conditions	If demand for energy falls, energy prices should decline. The level of the fall is determined by factors such as quantities of domestic energy supply, substitutability and market trading conditions. If energy prices fall, rebound effects could occur.
Economic output (change)	Trade balance	The changes in energy imports and exports as a result of changes in national energy consumption and prices as a result of either energy efficiency investment or energy demand reduction effects. The economic structure of a country will determine the trade flow as a result of energy efficiency.	■ Imports ■ Exports	For energy-importing countries reduced demand for energy can reduce energy import costs. For energy exporters, reduced energy demand frees up more supply for export if foreign demand for oil exceeds supply.* Another trade effect may appear if an economy becomes more competitive due to reduced energy costs, reducing export prices and increasing demand for that country's goods. However, trade effects may also be negative, if businesses become less competitive or if energy efficiency goods are imported.

* In the case of countries with energy subsidies, which are often energy exporters, the public sector bill funding the subsidies is reduced (see Chapter 3: Public Budget Impacts of Energy Efficiency).

© OECD/IEA, 2014.

Box 2.2 Transformational effects

A transformational effect due to energy efficiency measures can occur when "changes in technology ... change consumers' preferences, alter social institutions, and rearrange the organisation of production" (Greening et al., 2000). Technological developments may lead to both energy efficiency improvements and a different bundle of characteristics to the incumbent technology. In some cases, the characteristics may be more influential in driving economic growth than the efficiency improvements *per se*.

Several historical examples provide cases in point. In addition to improving efficiency, the shift from steam engines to electricity enabled much more flexible and decentralised production processes. Another example is the well-discussed Jevons paradox, in which Jevons (1865) put forward the idea that technological efficiency can increase the rate of consumption of a resource (in that case, coal) by reducing its unit cost. But Jevons argues this is a natural corollary to the increase of activity and wealth.

Because these effects involve a transformational shift in behaviour or technology, and also depend on a country's stage of economic development, they are very difficult to foresee – and are the least-documented of the macroeconomic impacts. It is difficult to predict transformational effects based on historical examples and to isolate the contribution of energy efficiency to a wider transformational effect. Nonetheless, it is worth considering which energy services have the potential to initiate a New Industrial Revolution, and what transformations in the global economy they might stimulate (Fouquet, 2014a).

This publication does not examine the transformational effect. It is anticipated that when assessing the multiple benefits of energy efficiency policies, policy makers will tend to estimate transformational effects in a qualitative (rather than quantitative) manner.

Sample results for key macroeconomic impacts

A summary of estimates of the four macroeconomic impacts (economic development measured by GDP, employment, energy price changes and trade balance) typically assessed in the energy efficiency and economics literature is given in this section to illustrate the scale of their importance.

Readers should be aware that comparing studies is difficult as they often differ in key aspects such as: (a) the methods used; (b) what is meant by (and included in) the efficiency improvement; (c) whether the spending on the efficiency-enhancing measures has been included in the estimation of the impacts; and (d) what sector(s) efficiency improvements occur in and the importance of these sectors in the context of the wider economy. This literature review aims only to highlight examples of estimates; the results shown should not be used for direct comparison.

Economic development, measured by GDP

Most macroeconomic modellers include GDP as an important output indicator of economic activity when looking at the impacts of energy efficiency measures on the economy.[9] GDP is unlikely to be influenced by energy efficiency measures when the relevant policy or

9 It is recognised that GDP may not be the best indicator of economic well-being. The World Bank and the United Nations Development programme suggest using the Human Development Index and other indices to reflect broader measures of economic well-being (http://hdr.undp.org/en/content/measuring-human-progress-21st-century). The Organisation for Economic Co-operation and Development (OECD) Better Life Index recognises that economic activity may not be the most relevant parameter to citizens and suggests other criteria (www.oecdbetterlifeindex.org).

© OECD/IEA, 2014.

programme affects only part of the economy (e.g. is sector-specific or small relative to the scale of the economy). Results from various samples below provide insights, with the usual caution in their interpretation due to differing metrics and scale of investments modelled.

An economic impact assessment of the (at the time) proposed Energy Efficiency Directive was carried out by Cambridge Econometrics for the European Union, using the E3ME macro-econometric model. It estimated that GDP could increase by a modest 0.25% if energy efficiency measures reduced PED by 15.4% (or 283 million tonnes of oil-equivalent [Mtoe]) by 2020 (compared with reference projections) and required additional investments of 0.84% (USD 36.3 billion or EUR 26.6 billion)[10] (EC, 2011).

Copenhagen Economics modelled the macroeconomic impacts of energy efficiency renovation of buildings in Europe. Energy efficiency measures delivering reductions of 65 Mtoe (5.4%) to 96 Mtoe (8.2%) in final energy demand in 2020 were estimated to require additional investment of USD 56 billion to 107 billion (EUR 41 billion to EUR 78 billion), resulting in an annual GDP increase of USD 209 billion to USD 397 billion (EUR 153 billion to EUR 291 billion). These values were calculated mainly by estimating the number of jobs created as a result of the investment in energy efficiency and using multipliers to estimate the aggregated gross value added per job in different sectors (Copenhagen Economics, 2012). The study found that this result could be achieved through regulatory reforms that removed structural barriers to energy efficiency; no new public subsidies were needed to deliver energy demand reduction for consumers and businesses.

At national level, the government of the United Kingdom commissioned two studies estimating macroeconomic impacts as part of a study on rebound effects associated with energy efficiency measures.[11] Econometric and computable general equilibrium (CGE) models indicated a reduction of final energy demand of 8% and 5% (respectively) in 2010 through energy efficiency measures, resulting in estimated GDP growth of 1.26% and 0.1% to 0.2% (respectively) (Barker and Foxon, 2008; Allan et al., 2006).

Investigation of the macroeconomic impacts of Germany's energy efficiency plan found that a final energy demand reduction of 6% by 2020 required annual investment of USD 16 billion (EUR 12 billion) in energy efficiency and delivered an increase in GDP of 0.7%. This estimation was also carried out using an econometric model (Lehr, Lutz and Edler, 2012).

Globally, the ENV-Linkages model (a CGE model) has been used by the OECD to estimate the worldwide economic impacts of achieving the Efficient World Scenario (EWS) described in the IEA *World Energy Outlook 2012* (IEA, 2012). Estimates revealed that reduction in PED of 6% in 2020 and 14% in 2035 would require, on average, an additional investment of USD 472 billion annually through to 2035. GDP would increase by 1.1% by 2035, with one-third of that occurring by 2020. This additional demand for jobs and investment may be positive in the current context, with many OECD member countries facing slack economies and unemployment levels above their structural level.

Employment

As supporting job creation is a major aim of policy makers, they often seek information about estimated employment effects resulting from energy efficiency programmes. In fact, net employment should be estimated, taking into account both employment gains and losses.

Establishing clear definitions of employment at the outset of the analysis is critical. Direct jobs are usually defined as those created in either manufacturing or installation of energy efficient

10 In 2000 currencies.

11 See discussion of rebound effect in Chapter 1. In estimating the macroeconomic rebound effects, researchers estimated the main macroeconomic impacts of energy efficiency measures and then examined whether these led to increased energy consumption.

© OECD/IEA, 2014.

equipment. Indirect jobs may result from supply chain effects. Combined, these provide the total gross employment impact. The net impact is determined by subtracting from this total any jobs that may be lost elsewhere in the economy, for example in energy production sectors, or as a result of limited labour market capacity leading to higher wage rates.

The effectiveness of an energy efficiency programme in creating jobs will depend on the size and structure of investment and the type of energy demand reduction interventions being supported. Compared with the same investment in the fossil fuel industry, energy efficiency services have been found to generate three times the number of jobs per million dollars invested (ACE, 2000; Pollin et al., 2009). Jobs in improving energy efficiency through the maintenance and repair of equipment and buildings can also be a significant source of employment and should be included. Energy efficiency jobs tend to range from low wage jobs to highly skilled technical work and are often geographically dispersed.

Many of the jobs (but by no means all) associated with energy efficiency are temporary in nature; thus, job years (i.e. the number of jobs multiplied by the number of years they last) rather than the number of jobs is a preferable unit of measurement for estimating employment impacts. An alternative is to provide an estimate of the job impact in one particular year. Policy makers can then use the term "jobs" (with appropriate caveats), as opposed to "job years", which is more difficult to communicate. Providing information on the skill levels or incomes associated with these jobs is important to helping policy makers understand what kinds of jobs are created.

Good sectoral analysis is needed to properly understand current spare labour capacity, skills available and changes in labour rates. Sector-specific methods are essential in detailing the jobs created in individual sectors and should then be aggregated to represent the whole economy. Economy-wide methods are especially useful in the context of labour market impacts, because jobs created in one sector may be partially offset by losses in another sector. A single-sector method might over- or underestimate the net jobs created.

While not specialised for labour markets, macro-econometric, input-output (I-O) and some CGE models can mimic realistic representations by estimating the number of jobs required, any labour market crowding out effects and whether capacity exists in the current labour market. What these models cannot assess is whether the available workforce has the necessary skills to fill the vacancies created by energy efficiency programmes. This requires a more specific assessment of the key sectors (mainly construction and engineering). If this supplementary analysis finds that skills shortages are likely, the modelling scenarios should be revised to take this into account as it could have macroeconomic implications (see example in Cambridge Econometrics, GHK and Warwick Institute for Employment Research, 2011). Comparison with individual department or ministry labour forecasts is useful, to check the validity of results.[12]

Most of the studies cited above for GDP have also estimated the potential for jobs created as a result of energy efficiency measures. Estimated values range from 7 to 22 job-years per EUR 1 million invested. Other authors report increases in employment per unit of reduced energy demand. The ENV-Linkages CGE model, similar to many CGE models, assumes a perfectly flexible labour market and thus calculates no net employment gains, instead estimating shifts of workers among sectors. A detailed description of how the jobs created through energy efficiency measures are estimated using an I-O model for Hungary along with an overview of the literature in this area is given by Urge-Vorsatz et al. (2010).

12 More information on approaches to estimating green jobs in general is available in the OECD *Greener Skills and Jobs* report (2014), www.oecd-ilibrary.org/docserver/download/8514041e.pdf?expires=1400600232&id=id&accname=ocid43019508&checksum=9352C84895921810CB73BC5D7BA6CCD3.

© OECD/IEA, 2014.

Price effects

Total primary energy demand (PED) and the energy mix are important factors to consider in predicting energy prices. In addition to reducing overall demand, improvements in energy efficiency may change consumption of energy in one fuel more than others – either because of the sector affected or because of energy price differences across fuels – with resulting implications for fuel prices and overall energy prices. As energy fuels are globally traded commodities, unless energy efficiency measures are implemented at a large enough scale, it is unlikely that global energy prices will change significantly. Should the trend to improve energy efficiency continue at a global scale, however, energy demand growth should be at least tempered, with the effect of driving energy prices lower than they otherwise would have been.

The macroeconomic implications of larger changes in energy prices could be huge. The oil price spike of the early 1970s was a key factor in the economic crisis at the time.[13] By contrast, the financial recession of 2008 caused demand for energy to contract, and oil prices fell from a high of USD 147 per barrel (USD/bbl) in July 2008 to a low of USD 32/bbl in December 2008 (by 2012, the average annual price had risen again to USD 108/bbl, a record price in real terms). The European Climate Foundation roadmap analysis found that a doubling in oil prices for three years could cost the European Union (EU) economy USD 410 billion (EUR 300 billion) over the same time period (ECF, 2010).

Different energy efficiency models take different approaches to estimating energy prices. Energy prices can be calculated exogenously and then included as part of the inputs; conversely, energy prices can also form one of the outputs that occur as a result of the energy efficiency measures. In its EWS, the IEA *World Energy Outlook 2012* uses a different method for modelling energy prices. Whereas most *World Energy Outlook* scenarios set energy prices as part of the scenario inputs, the EWS sets policy measures and allows energy prices to change to reflect the change in equilibrium between energy demand and supply. The EWS estimates that a 6% reduction in global PED through energy efficiency measures in 2020, following on to 14% in 2035, would drive crude oil prices down by USD 16/bbl in 2035 compared with the New Policies Scenario.

Trade balance

Trade balances, i.e. a country's exports net of imports, are included in the GDP calculations above but warrant additional exploration to better understand the phenomenon as it relates to energy efficiency measures.

To date, results from studies modelling the macroeconomic impacts of energy efficiency measures are divided on whether trade balances are positively or negatively affected. This disparity arises mainly from differences in what is actually being modelled. Of the macroeconomic studies mentioned to estimate GDP, ENV-Linkages (a global model) finds that world trade increases marginally (0.2%) through strong stimulation of less energy-intensive goods and services in OECD member countries in particular, whereas energy transformation industries experience a slowdown in activity and trade (up to 8.9% decrease).

Energy efficiency measures can drive an increase in both exports and imports. For example, a national-level study in Germany showed that increased demand for energy efficiency goods and services creates a consumer surplus that drives up private consumption, initially leading to increased imports. The impacts are positive for trade: energy imports decline while exports of energy efficiency goods and services increase – if early initiative is taken.

13 Not all the implications are negative; high energy prices can also stimulate innovation and reduce demand for energy.

© OECD/IEA, 2014.

Energy-exporting countries are likely to suffer an export loss if energy efficiency measures are implemented on a large scale. Export volumes could be expected to decline in countries producing at the margin, while export earnings are likely to decline in all energy-exporting countries if the demand reduction is large enough to drive down energy prices. While the global ENV-Linkages model shows more export trade globally, it also reveals that increased implementation of energy efficiency measures is likely to result in winners and losers in the global trade of energy.

Energy imports, by contrast, could be reduced in energy-importing countries by energy efficiency measures. One study estimates that Germany's gas import bill from Russia could be halved in ten years through energy efficiency measures in the industrial and buildings sectors (Ecofys, 2014).

Methodological approaches

Diverse tools make it possible to estimate the macroeconomic impact of energy efficiency measures as part of a policy appraisal process,[14] which should include an assessment of the range of public and private costs and the likely impacts (see the Companion Guide at the end of this publication). To the greatest degree possible, these impacts should be quantified and monetised as part of a benefit-cost analysis (BCA).[15] If this is not possible, qualitative analysis should be used to account for impacts that cannot be monetised. This allows policy makers and the public, once results are published, to understand the rationale and consequences for implementing a policy. A decision tree provided in the Companion Guide at the end of this publication is a useful tool for selecting the assessment method most suitable for the circumstances and requirements.

Methodological options and indicators

Quantitative estimation of economy-wide economic impacts of energy efficiency provides a means to understand and compare against other measures in terms of costs, benefits and scale. Basic analysis can provide some important insights without high resource requirements. But the complex nature of interactions that a change in energy efficiency policy – and the related large-scale investment – can prompt across the economy usually requires more complex computer models that simulate different energy efficiency or economic scenarios.

Basic analysis of the macroeconomic impacts of energy efficiency can be carried out using a spreadsheet; the scope of the analysis can be set as wide or as narrow as desired. Initially, a few key estimates are needed such as the scale of the costs of the programme, the likely investments needed, the value of the energy demand reduction and the sectors in which increased spending is likely to occur. A simple calculation can then be carried out to estimate key macroeconomic variables (e.g. consumer spending, investment, net exports and government spending) with and without the energy efficiency programme.

This calculation might be carried out for a single sector or several sectors that could be aggregated to estimate GDP. Such basic analysis is most likely to be useful to make a first estimate of the direct effects of the programme. If values for multipliers are available from other models that show, for example, the indirect effects of spending on energy efficiency, these can also be used within a basic assessment.

14 It could also be part of an *ex post* policy evaluation using historical data.

15 Although the commonly used term is "cost-benefit analysis" (CBA), the IEA prefers the term "benefit-cost analysis" (BCA) due to the fact that the ratios produced are expressed as "benefit:cost". The actual approach is the same.

© OECD/IEA, 2014

More advanced modelling to assess the macroeconomic impacts of energy efficiency measures can be carried out using a number of different methods. CGE and macro-econometric models are the main methods used, but are often combined with others, such as I-O tables and analysis. An important point is that macroeconomic models generally represent sector details, but do not support technology details.

All CGE and econometric models should be capable of producing the same type of results, including some or all of the following: GDP, employment, trade, energy prices and expenditures, investment, consumer price index (CPI), value-added, distribution, public budget, and activity levels in social and economic spheres. Policy makers should be aware of commonly used models, as well as their characteristics, scope and impacts (Table 2.2).

Table 2.2 Overview of several macroeconomic models of energy efficiency impacts

Model name	Model type	Scope	Impacts
World Energy Model[A]	Partial equilibrium	Global	Energy prices and expenditures, investment
GINFORS[B]	Econometric	Global	GDP, employment, trade, CPI, distribution
E3ME[C]	Econometric	EU member states	GDP, employment, trade, CPI, distribution
ENV-Linkages[D]	CGE	Global	GDP, employment, trade and value-added by sector
ThreeME[E]	CGE	France	GDP, employment, trade, distribution, public budget
HMRC CGE model[F]	CGE and BCA	United Kingdom	GDP, employment, public budget
REMI[G]	CGE and I-O	Canadian provinces	GDP, employment, public budget
UKENVI[H]	CGE and I-O	United Kingdom	GDP, employment, trade, public budget, aggregate distribution effects, investment behaviour and sectoral activity levels
IKARIS[I]	Bottom-up buildings systems model with I-O	Germany	Public budgets, employment
3CSEP model[J]	Bottom-up buildings sector with I-O	Hungary	GDP, employment
Copenhagen Economics model[K]	PCGE/macroeconomic multipliers	Regional (EU)	GDP, employment, trade, CPI
PANTA RHEI[L]	I-O	Germany	Employment, trade, value-added, production
SEAI model[M]	BCA	Ireland	GDP, employment, public budget

Notes: PCGE = partial computable general equilibrium; SEAI = Sustainable Energy Authority of Ireland.
The models included in this table are those which were presented at the IEA Roundtable on Macroeconomic Impacts of Energy Efficiency, Paris, January 2013. This list does not include all macroeconomic models, but is intended to provide a comprehensive sample.
Sources: A. IEA (2013b and 2012); B. Lehr, Lutz and Ulrich (2013); C. Cambridge Econometrics (2014); D. Château, Magné and Cozzi (2014); E. Callonnec et al., (2013); F. UK HMRC (2013); G. ENE (2012); H. Allan et al., (2006); I. Kronenberg, Kuckshinrichs and Hansen (2012); J. Urge-Vorsatz et al. (2010); K. Copenhagen Economics (2012); L. Lehr, Lutz and Edler (2012); M. Scheer and Motherway (2011).

Incorporating both bottom-up and top-down methods enhances the results when modelling the macroeconomic impacts of energy efficiency measures. This allows engineering knowledge (bottom-up) in the relevant sectors to be combined with economists' expertise (top-down) on the wider economy. Moreover, a bottom-up model provides a useful tool to

© OECD/IEA, 2014.

develop a strong understanding of the microeconomic underpinnings of a sector, which can be combined with a top-down model to achieve economy-wide coverage. Data, time and financial resource constraints may make it difficult for policy makers to carry out this detailed analysis (Ghersi and Hourcade, 2006; Hourcade et al., 2006).

In general, it is unlikely that one single model will be able to model everything well; policy makers can expect that several purpose-built models will be used together to model economy-wide effects.

Steps in modelling macroeconomic impacts

The IEA Policy Pathway series outlines the key steps in the energy efficiency policy-making process; the multiple benefits approach has implications for each of these steps (See Table 1 in the Companion Guide at the end of this publication). Assessing macroeconomic impacts requires an intensive effort at the policy planning stages, when various policy options can be appraised based on the impacts they might be expected to have. In order to give policy makers a sense of how the macroeconomic assessment could be carried out in practice, a possible step-by-step process for policy makers modelling the macroeconomic effects is outlined below, noting the need to take into account different situations in terms of data, expertise and other resources (Box 2.3). Broadly, the sequence would begin with an initial assessment of the macroeconomic impacts to determine whether they are significant enough to warrant further investigation. If the answer is positive, a detailed analysis can be pursued.

Box 2.3 Possible steps in an assessment of macroeconomic impacts of energy efficiency measures

Plan

- Identify which macroeconomic indicators should be estimated in addition to energy demand reduction of the policy.

Estimate

- Estimate the energy demand reduction as a result of the policy (this should be routinely carried out as part of an impact assessment of an energy efficiency policy measure but is repeated here for emphasis).
- Carry out a basic assessment of the macro-economic impacts.
- Assess whether economy-wide impacts are significant enough to merit detailed estimation.
- Select the method for detailed assessment.
- Estimate of the macroeconomic impacts of energy efficiency policy scenarios, including baseline.

Verify

- Consider whether all issues are included.
- Conduct a sensitivity analysis.

Plan

- **Identify which macroeconomic indicators should be estimated**

Assessment of a wide range and number of macroeconomic impacts can quickly become overly complex; thus, it is important at the outset to establish some priority indicators for examination. These indicators could be ranked in order of priority to allow for initial examination, adding others if more detailed analysis is desired.

© OECD/IEA, 2014.

Different stakeholders will place different value on macroeconomic indicators. Finance ministries generally consider changes to GDP, trade balance, levels of employment and public budgets most important. Energy ministries might be more concerned with energy security. Industry and the producing sectors are likely to be interested in levels of investment and any impacts on profits that might arise as a result of fiscal policies, increased demand for goods and services they can provide, or changes in energy prices. The general public is likely to favour changes to rates of employment, energy prices, disposable income, the CPI and any distributional effects.

Based on the choice and feasibility indicated by the policy and academic literature,[16] six main macroeconomic impacts are most likely to be requested by policy makers as first and second priorities (Table 2.3).

Table 2.3 Macroeconomic impacts included in energy efficiency policy appraisal

1st priority	2nd priority
GDP	Trade balance
Employment	Energy prices
Disposable income	Energy security

Estimate

Estimate the energy demand reduction as a result of the policy

A critical first calculation for estimating macroeconomic (or any other) effects from energy efficiency measures is likely to be the amount of energy demand reduction delivered by each policy option compared with the baseline. This is because the key drivers of macroeconomic impacts (described above) depend on a reduction in energy costs as a result of the energy efficiency measures. Many of the multiple benefits or impacts of the measures are linked to the energy demand reduction actually achieved (see Chapter 1).

It is important to consider the timescale over which the estimation of effects and impacts should be carried out, as it may be critical in demonstrating the full macroeconomic impacts of energy efficiency measures. The up-front, short-term investment costs may be offset against longer-term energy demand reduction effects. In the longer term, the effect of energy efficiency measures should become positive as the energy demand reductions accumulate. The length of time required to do this generally depends on the sector, the amount of energy reduced and the technology cost.

Carry out a basic assessment of the macroeconomic impacts

Simple calculations with a spreadsheet can provide a first assessment of whether the potential impact(s) from energy efficiency measures are significant enough to warrant more detailed modelling. Although some data are required, this basic assessment does not generally involve sophisticated analysis or much time. This early stage assessment gives an important opportunity to examine what data are available for more detailed assessment.

Simple equations can be used to estimate the direct macroeconomic effects of energy efficiency measures, but feedback mechanisms are not usually included in such equations.

16 See presentations from Roundtable on the Macroeconomic Impacts of Energy Efficiency Improvements, 24-25 January 2013: www.iea.org/workshop/roundtableonthemacroeconomicimpactsofenergyefficiencyimprovements.html.

© OECD/IEA, 2014.

It should be noted that the direct impact on GDP calculated is not likely to be large because of the absence of indirect impacts and because energy expenses are a limited share of total GDP.[17] Since indirect impacts contribute significantly to the overall macroeconomic impacts, multiplier factors may be available in many countries to enable a rough calculation of the potential change in sectoral employment for every dollar spent in different sectors and the resulting change in GDP.

The calculated estimates can focus on the first year of the policy and can then be projected forward using official economic projections to assess the cumulative effect of the measures into the future.[18] The economic parameters used should match those included in the baseline projections, against which the policy scenarios will be compared. It should be noted that the very rough estimates that this process generates should be used only as a rough guide, even if the decision is made not to proceed with more detailed estimation.

- **Assess whether economy-wide impacts are significant enough to merit detailed estimation**

Some analysis of the basic assessment results is usually needed to support a decision on whether more detailed estimation of the macroeconomic effects is worthwhile. Macroeconomists and ministry of finance officials can be consulted for their views on the level at which they would consider a macroeconomic indicator significant and worthy of further investigation.

A threshold indicator for changes to GDP or employment, for example, can be chosen, above which more detailed modelling could be carried out. Modelling results suggest that when the impacts indicate more than 0.5% change, it becomes more worthwhile to carry out detailed analysis of the macroeconomic impacts of the energy efficiency measures. Nevertheless, this figure should not prevent detailed modelling analysis where it is merited for other reasons. For example, while only a small change in net employment may be estimated from the initial basic assessment, the shift of employment between sectors may be considered significant and merit further investigation. In general, if the energy demand reduction delivered by the measure is low, second-round effects of the energy efficiency measures throughout the economy are likely to be small.

- **Select the method for detailed assessment**

Numerous factors can come into play when deciding how to go about estimating a set of macroeconomic impacts of energy efficiency measures, including the quality of data available, modelling resources, time constraints and the type of impacts expected. Calculating the macroeconomic impacts is more likely to require cross-sectoral methods such as CGE or macro-econometric models, particularly given the need to include indirect or second-round effects in the estimation. The Companion Guide at the end of this publication outlines the different modelling methods available and includes a decision tree that can be used to select the most appropriate modelling method.

- **Estimate the macroeconomic impacts of energy efficiency policy scenarios, including baseline**

A robust estimation process begins with collecting data to provide input parameters for all the variables required in the model. It is critical to establish a baseline (or BAU) situation

17 Any energy bill savings are likely to be redirected into spending on other goods and services or savings that are then lent to businesses by banks. There may be a shift in the way income is spent, from energy to other goods and services, but the absolute amount spent remains the same. It is more likely that this shift in consumer spending will have indirect impacts such as changes in employment and government budgets that affect GDP.

18 Data required are likely to include: official estimates of current and projected GDP, population, energy demand, employment and trade balance.

© OECD/IEA, 2014.

against which the different policy scenarios can be compared. Sensitivity analysis can then be carried out to test the robustness of the baseline. The policy scenarios can then be expressed as parameters and modelled using the same underlying economic projections to allow comparison of how various options for energy efficiency policy influence macroeconomic impacts.

Verify

- **Consider whether all issues are included**

As a verification measure, it is recommended to consider whether the special policy-making considerations outlined below have been included in the modelling method. For example, how are rebound effects treated in the modelling estimations? The modeller and policy makers should consider the assumptions in the model with regard to crowding out. As distributional effects are often a key concern of politicians, the model can be useful in providing some detail on the societal distribution of the macroeconomic impacts, including the identification of winners and losers as a result of the policy measures.

- **Conduct a sensitivity analysis**

Running a sensitivity analysis provides a means to test the robustness of the policy assessment. This involves changing some of the input variables, such as energy prices or the underlying economic variables, and rerunning the model to see how these changes alter the results. If a small change in some of the parameters dramatically alters the results, closer examination may be needed of the input variables and the relationships underpinning the model. A sensitivity analysis can also be extended to the baseline to test whether estimations based on different baseline yield consistent results.

Default values and approximation methods

When resource or time constraints make it impossible to undertake in-depth analysis of the macroeconomic benefits of energy efficiency measures, a range of sample values derived from the estimation results of other methods can serve as proxies. A disclaimer should be added that it is very difficult to compare study results without knowing in detail the assumptions made (such as time horizon, likelihood of crowding out, region of analysis, method of financing the energy efficiency measures, or the economic values of the energy demand reduction using prices of gas, oil and coal). Sample values from the literature may allow policy makers to make "back-of-the-envelope" estimates of the macroeconomic impacts expected from various energy efficiency measures (Table 2.4).[19]

The variety of metrics used makes it challenging to compare results among models. Indeed the most common metric for macroeconomic effects is percentages, but these do not allow for a full comparison among models since simply comparing percentage increase in, for example, GDP does not capture the size of the economy, the level of investment required or the energy demand reduction delivered. For this reason, this publication aims to show some key macroeconomic impacts – GDP, job creation, household consumption – relative to both unit investment and the PED savings. Data available from comparable studies were insufficient to be able to provide estimates for energy price and trade balance impacts.

The level of energy demand reduction achieved (e.g. % savings) will impact GDP and jobs metrics. When modelling a range of annual savings targets over a 15-year period, for example, Environment Northeast found that as the level of effort increases, the macroeconomic impacts decline because higher energy demand reduction typically requires

19 The authors recognise that some of the results are quite heterogeneous as a result of these different factors; policy makers are advised to examine the studies behind these values for more detail on their applicability.

© OECD/IEA, 2014.

more expensive efficiency measures. Establishing a range of impact estimates based on various levels of effort allows policy makers to choose the multiplier that best corresponds with the energy demand reduction expected from existing or planned programmes. The fuel type (i.e. avoided energy cost) will also have an important influence. As shown in the New England and Eastern Canada studies, the energy efficiency measures (in buildings) had a significantly greater impact on heating oil consumption, for the most part due to the relatively high price of oil and the type of energy efficiency measures implemented (ENE, 2013).

Table 2.4 A selection of estimates of macroeconomic impacts of energy efficiency programmes

	Range	Mean	Median	Studies
Amount invested per PED savings (EUR billion/Mtoe)	0.09-0.63	0.51	0.45	Copenhagen Economics, 2012; EC, 2011; Lehr, Lutz and Edler, 2012; OECD, 2013
Change in GDP per unit of investment (EUR/EUR)*	0.91-3.73	1.31	1.81	Copenhagen Economics, 2012; EC, 2011; Lehr, Lutz and Edler, 2012
Jobs created per year per unit investment (jobs/EUR million)**	9.2-17.07	9.95	11.64	Copenhagen Economics, 2012; EC, 2011; Lehr, Lutz and Edler, 2012
Jobs created per PED savings (jobs per ktoe)**	0.76-19.61	0.92	7.06	Copenhagen Economics, 2012; Lehr, Lutz and Edler, 2012; Barker and Foxon, 2008
Change in household income per investment (EUR/EUR)***	-0.16-0.88	0.32	0.34	OECD, 2013; Lehr, Lutz and Edler, 2012; EC, 2011

Note: EUR/Mtoe = euros per million tonnes of oil-equivalent; ktoe = thousand tonnes of oil-equivalent.

* This value is an economic multiplier and would therefore be highly dependent on the usual multiplier assumptions, such as the size of the area involved and its trade ratios.

** The largest variation in the modelled estimates relates to job creation where the numbers of jobs created per unit of energy and investment varies hugely. This is likely to result from definitional issues, since jobs figures need to be defined as direct, total, gross or net, and by the nature of the job units, which also vary significantly by sector.

*** Household income change estimations involve very many assumptions, particularly regarding whether the money saved from energy bills is spent or saved. It is therefore questionable whether to include these values in Table 2.4.

The results reported in Table 2.4 should be interpreted with caution; for most impacts, only three studies are included in the calculation. It was not possible to calculate all indicators from all of the main studies sourced because of the difficulty in interpreting the data. Also, some variation is usually evident in the results by sector. However, the results should provide some "ballpark" figures for policy makers wishing to estimate the macroeconomic impacts of an energy efficiency programme. For a given amount of energy demand reduction or investment in energy efficiency measures, the table can be used to estimate the GDP, job creation and household consumption impacts. By way of sensitivity analysis, it is recommended that the upper and lower values in the ranges be used to calculate a higher and lower bound for the estimation.

Policy-making considerations

Certain issues must be addressed when assessing the macroeconomic impacts of energy efficiency measures. Some are general considerations associated with assessing macroeconomic effects and others more related specifically to energy efficiency impacts; they also depend on the macroeconomic modelling method chosen, if any. Two aspects of

© OECD/IEA, 2014.

particular relevance warrant further investigation: the effects of energy price and crowding out. As in other chapters of this book, the rebound effect is also examined in context.

Energy price assumptions

Energy prices may be significant macroeconomic impacts of energy efficiency measures but they are also very important parameters in modelling the impact of energy efficiency improvements on the economy. Thus, careful attention is needed to how energy prices are treated. While energy efficiency improvements can reduce energy prices, changes to energy prices also drive energy efficiency technology adoption rates and hence drive down energy demand (Popp, Newell and Jaffe, 2009; Copenhagen Economics, 2010).

Economic models use price elasticities to represent how people and firms respond to changes in the price of energy and indeed all goods included in the macroeconomic model. Price elasticities can be estimated in bottom-up models and then input to the macroeconomic mode, and have been used extensively in the literature (Gillingham et al., 2009). Policy makers need to be aware of the key price elasticities and assumptions included in the model, as these can be an important determinant of results. For example, *should the same price elasticities be assumed for a change in price that results from taxes compared to a fluctuating energy price?* Also, price elasticities are not the same at all price levels and depend on the time frame. Stakeholder responses to price increases may vary depending on how the change is applied and the type of investment.

Ideally, models should allow energy prices to rise and fall as energy efficiency improves and energy consumption declines. Such a dynamic modelling method introduces a mechanism to partially account for or represent the rebound effect, as falling energy prices often prompt direct take-back in the form of higher energy consumption (Rebound effect perspective 2). Price effects are uncertain and thus excluded from many models. They are, however, very important in relation to estimating rebound and economic effects of energy efficiency policies and future modelling of energy efficiency impacts should endeavour to be more dynamic.

Models estimating macroeconomic impacts of energy efficiency measures should be able to account for regional energy sources and prices, as policy choices are inextricably linked to the type and price of energy fuels. Current models dealing with energy efficiency policy do not sufficiently take into account choices regarding energy supply issues, such as distributed generation and storage. As these technologies are increasingly deployed, it will be necessary to begin accommodating them within top-down or bottom-up economic models.

Finally, the duration of energy price changes strongly influences behaviour, especially if revenues and return to capital changes, reducing the incentives to invest or to replace capital/stock capacity. This effect should also be accounted for in models. Several studies show that the long-term effects of higher energy prices or taxes may exceed the short-term effects by a factor of 3 to 4. This reflects the highly capital-intensive nature of energy services (heating, transportation, etc.) that account for most of the activities characterised by high energy intensity, which results in high inertia in purchasing behaviour. Consumers of these services are less likely to respond to short-term changes in price. Many models have problems in accounting for these temporal nuances.

Crowding out

Spending on energy efficiency can "crowd out" (or "crowd in") other investments. The greater the scale and impact of energy efficiency policy, the more relevant the crowding out phenomenon becomes. Since policy makers seek to encourage more, not less, private investment, it would be very helpful if macroeconomic models could examine this effect in some detail.

© OECD/IEA, 2014.

Rebound effect perspective 2

The macroeconomic rebound effect

Some of the macroeconomic impacts from energy efficiency measures outlined above can lead to a rise in energy consumption relative to the energy demand reduction estimated based on the measure's technical potential. Known as the macroeconomic rebound effect, such an outcome has generally been viewed negatively. This chapter recognises, in contrast, that the rebound effect may have positive societal welfare impacts that may be as – or even more – important for policy makers than the reduction in energy demand. Ergo, rebound is not always negative.

Rebound effects should, of course, be included in benefit-cost calculations of energy efficiency measures; thus, it is very important to also include the welfare effects to avoid bias. The rebound effect must also be understood against the backdrop of the particular energy efficiency policy and the economic conditions: i.e. the rebound effect is time-, space-, policy-, economy- and sector-specific. Time is particularly important as the macroeconomic rebound effect may have long-run outcomes that only become visible over time.

The macroeconomic rebound effect is different for the producing and consuming sectors as their respective increases in energy consumption arise through different processes (Turner, 2013). It is also important to understand rebound effect in an action context, i.e. not just that it exists, but what policy design measures or packages could be used to offset the effect.

Macroeconomic or top-down models are a powerful tool in conceptualising and estimating the economy-wide rebound effects (both direct and indirect) of specific energy efficiency policy measures. The direct rebound effect is likely best estimated in a sectoral, bottom-up model from which the results can then be fed into a macro model. The indirect rebound effect, which has been less estimated, has implications for the macroeconomic effects of energy efficiency measures since the two are inextricably linked. Recent research suggests that the indirect rebound effects can be negative in some cases (Turner, 2013).

The total macroeconomic rebound effect is estimated to be in the range 10% to 30% in the United Kingdom, which is perhaps representative of other developed countries (Sorrell, 2007). The effect can be expected to be higher in developing countries where growth rates are high and a greater concentration of more energy-intensive activities is evident. Others estimate that, for energy efficiency measures undertaken globally in the period 2013-20, the total global average economy-wide rebound could amount to 31% of the projected energy demand reduction potential by 2020, and rise to 52% by 2030 (Barker, Dagoumas and Rubin, 2009). As this is a global average, the macroeconomic rebound effect is likely to be higher than 50% in some countries (some experts, however, dispute this result).

The degree of crowding out that might occur is not adequately addressed by existing models (Pollitt, 2014). Few models include crowding out (or crowding in) effects where increased need for investment in energy demand reduction would make less room for other activities, including by driving up inflation and demand for labour. Crowding out can be implicitly modelled; for example, inflation and labour demand can increase. Other channels of crowding out, such as an increase in interest rates or investment competition within a company if the investment budget is fixed, may not be considered (Lehr, Lutz and Edler, 2012). If increased investment is financed through increased taxes, then crowding out may not be an issue, as this leads to increased savings. If, however, investment is financed from other sources and access to capital is tight, crowding out could occur.

CGE models assume that all capital is allocated optimally, so an increase in investment in energy efficiency will mean that resources must be diverted from elsewhere (i.e. crowding

© OECD/IEA, 2014.

out). Not all econometric models incorporate crowding out as a standard feature (CGE models generally do).[20]

In reality, the speed at which crowding out occurs depends on the structural and economic nature of a country, including its position in the business cycle, the level to which the financial system is developed and the scale of the investment in energy efficiency. It is important to compare the impact of energy efficiency investments with other investment choices. A counterfactual scenario is needed to determine whether the energy efficiency investments would have happened without the energy efficiency policies, i.e. whether there is additionally associated with the policy measures or whether private investment has been crowded out. There may also be a problem of double-counting and/or free-riding; many models include some assumptions on free-ridership, but it is difficult to verify their accuracy.

Further research for stakeholders

Although understanding of the macroeconomic impacts of energy efficiency is advancing fast, several aspects remain challenging and merit focused attention by researchers and other stakeholders in the short term (Table 2.5).

Table 2.5 Further stakeholder research and collaboration opportunities in macroeconomic impacts

Area	Specific actions
Benefit areas and causal linkages	More work is needed to better understand the range of impacts that can occur with regard to spending, trade balance and energy price effects at national and global levels. Improve employment estimations to better account for shifts in demand between sectors; linkage to specialised labour market models may be useful. More study of producing sectors and macroeconomic impacts of energy efficiency in those sectors is needed.
Data, indicators and metrics	Identify key relevant macroeconomic indicators, other than GDP, that reflect outcomes for energy efficiency measures, drawing from sources such as the Human Development Index. Develop more robust and comparable metrics with transparent assumptions to allow better comparison of the results of macroeconomic modelling of energy efficiency measures.
Assessment methodologies	Better linkage of bottom-up engineering and top-down macroeconomic models is needed. Develop methods for integrating price effects into modelling to better estimating rebound and economic effects of energy efficiency policies in a more dynamic way. Improve treatment of energy efficiency market failures such as split incentives in macroeconomic models. Integrate crowding out into modelling in a more systematic way.
Collaborative initiatives	Seek opportunities for macroeconomic modellers to share experience and discuss the basis of assumptions used in their models, in order to generate more alignment. In governments, more collaboration between ministries of finance and energy would be helpful to improve modelling the macroeconomic impacts (usually the responsibility of the ministry of finance) of energy efficiency measures (usually originating in the ministry of energy).

20 In CGE models, new investment means potential for total capital stock to increase, rather than just being reallocated among sectors (leading to crowding out). CGE models may then show short-run constraints on total capital, but once investment kicks in (through different dynamic processes, rather than traditional static CGE), this no longer needs to be the outcome.

© OECD/IEA, 2014.

Conclusions

Macroeconomic impacts are an important element of assessing energy efficiency policies and the measures they include. In fact, these impacts may even be of more interest to budget-wielding ministries, such as ministries of finance, than the more traditionally promoted energy or environmental impacts. The macroeconomic impacts of energy efficiency programmes are generally positive in that programmes usually lead to increased economic activity, either directly or indirectly.

Robust assessment of the macroeconomic impacts is essential to provide credible information to policy makers, who may not be familiar with energy efficiency measures and targets. Aggregate, positive macroeconomic impacts may, for example, mask potentially contrasting distributional effects that should be investigated.

Macroeconomic assessment can be data-intensive, requiring significant expertise and software. Time and resources should be allocated accordingly, and the method selected should be tailored to the resources of the public institution.

Increased economic activity is likely to temper the energy demand reduction from the technical potential initially indicated for an energy efficiency programme; in other words, a macroeconomic rebound effect may occur. As positive societal and welfare benefits may be at the root of the macroeconomic rebound effect, the economic gain should be balanced politically against some expected decrease in the energy demand reduction or emissions saved. It is possible that the future economy may be driven by cost-saving efficiency gains, rather than the current growth model of consumer spending to exploit cheap energy, as has been the case for the past two centuries. More analysis is needed of these effects.

Several realistic messages for policy makers flow out of this analysis. Clearly, energy efficiency measures cannot be expected to solve a nation's economic woes, or to provide the complete solution to social and environmental problems. The first and foremost goal of energy efficiency programmes should be to reduce energy demand at least cost. But the macroeconomic benefits demonstrate that such programmes deliver high value across the economy. A broader calculation should dispel any thinking that energy efficiency programmes are a burden to the economy.

Bibliography

ACE (Association for the Conservation of Energy) (2000), *Energy Efficiency and Jobs: UK Issues and Case Studies*, report to the Energy Savings Trust, ACE, London, www.ukace.org/wp-content/uploads/2012/11/ACE-Research-2000-09-Energy-Efficiency-and-Jobs-UK-Issues-and-Case-Studies-Case-Studies.pdf (accessed 23 June 2014).

Allan, G. et al. (2006), *The Macroeconomic Rebound Effect and the UK Economy*, Final Report to the Department of Environment Food and Rural Affairs, University of Strathclyde, Glasgow, http://ukerc.rl.ac.uk/pdf/ee01015_final_a.pdf (accessed 23 June 2014).

Barker, T., A. Dagoumas and J. Rubin (2009), "The macroeconomic rebound effect and the world economy", *Energy Efficiency*, Vol. 2, No. 4, Springer Netherlands, Houten, pp. 411-427, www.unternehmenssteuertag.de/fileadmin/user_upload/Redaktion/Seco@home/Projektpartner_Ergebnisse/macroeconomicRebound.pdf (accessed 23 June 2014).

Barker, T. and T. Foxon (2008), *The Macroeconomic Rebound Effect and the UK Economy*, Research Report REF UKERC/WP/ESM/2008/001, 5 February 2008, UKERC (UK Energy Research Centre), London, http://bit.ly/1lLpDjB (accessed 23 June 2014).

Barker, T., P. Ekins and T. Foxon (2007), "The macroeconomic rebound effect and the UK economy", *Energy Policy*, Vol. 35, No. 10, Elsevier Ltd., Amsterdam, pp. 4935-4946, www.sciencedirect.com/science/article/pii/S0301421507001565 (accessed 23 June 2014).

© OECD/IEA, 2014.

Callonec, G. et al. (2013), *A Full Description of the Three-ME Model: Multi-sector Macroeconomic Model for the Evaluation of Environmental and Energy Policy*, ADEME (Agence de l'environnement et de la maîtrise de l'énergie) and OFCE (Observatoire français des conjonctures économiques), Paris, www.ofce.sciences-po.fr/pdf/documents/threeme/doc1.pdf (accessed 23 June 2014).

Cambridge Econometrics (2014), *E3ME Technical Manual, Version 6.0.*, Cambridge Econometrics, Cambridge, www.camecon.com/Libraries/Downloadable_Files/E3ME_Manual.sflb.ashx (accessed 23 June 2013).

Cambridge Econometrics, GHK and Warwick Institute for Employment Research (2011), *Studies on Sustainability Issues – Green Jobs; Trade and Labour*, Final Report to the European Commission, Directorate-General for Employment, 11 April 2011, Cambridge Econometrics, Cambridge, http://ec.europa.eu/social/BlobServlet?docId=7436&langId=en (accessed 23 June 2014).

Chateau, J., B. Magné and L. Cozzi (2014), "Economic Implications of the IEA Efficient World Scenario", *OECD Environment Working Papers*, No. 64, OECD Publishing, Paris.

Cherp, A. and J. Jewell (2011), "The three perspectives on energy security: intellectual history, disciplinary roots and the potential for integration", *Current Opinion in Environmental Sustainability*, Vol. 3, No. 4, Elsevier B.V., Amsterdam, pp. 202-212, www.sciencedirect.com/science/article/pii/S1877343511000583 (accessed 23 June 2014).

Copenhagen Economics (2012), *Multiple Benefits of Investing in Energy-efficient Renovation of Buildings – Impact on Finances*, report commissioned by Renovate Europe, Renovate Europe, Brussels, www.renovate-europe.eu/uploads/Multiple%20benefits%20of%20EE%20renovations%20in%20buildings%20-%20Full%20report%20and%20appendix.pdf (accessed 23 June 2014).

Ecofys (2014), *Energieabhaengigkeit von Russland durch Energie-Effizienz reduzieren (Reduce energy dependence on Russia through energy efficiency)*, report for DENEFF (Deutsche Unternehmensinitiative Energieeffizienz e.V.) (German enterprise energy efficiency initiative), Project number DESDE14880, Ecofys, Utrecht, www.deneff.org/fileadmin/user_upload/20141416_Ecofys_Studie_-_Energieabhaengigkeit_von_Russland_durch_Energieeffizienz_reduzieren.pdf (accessed 23 June 2014).

EC (European Commission) (2011), Commission Staff Working Paper Impact Assessment, Accompanying the document Directive of the European Parliament and of the Council on Energy Efficiency and amending and subsequently repealing Directives 2004/8/EC and 2006/32/EC, SEC(2011) 779 final, Brussels, 22.6.2011, EC, Brussels, http://ec.europa.eu/energy/efficiency/eed/doc/2011_directive/sec_2011_0779_impact_assessment.pdf (accessed 22 June 2014).

ECF (European Climate Foundation) (2010), *Roadmap 2050: A Practical Guide to a Prosperous Low Carbon Europe: A Technical Analysis*, ECF, Brussels, www.roadmap2050.eu/attachments/files/Volume1_fullreport_PressPack.pdf (accessed 23 June 2014).

EEA (European Environmental Agency) (2011), *Environmental Tax Reform in Europe: Opportunities for Eco-innovation*, EEA Technical report 17/2011, EEA, Copenhagen, www.eea.europa.eu/publications/environmental-tax-reform-opportunities (accessed 23 June 2014).

ENE (Environment Northeast) (2012), *Energy Efficiency: Engine of Economic Growth in Eastern Canada*, ENE, Ottawa, www.env-ne.org/public/resources/ENE_EnergyEfficiencyEngineofEconomicGrowth_EasternCanada_EN_2012_0611_FINAL.pdf (accessed 23 June 2014).

Fouquet, R. (2014a), Personal communication.

Fouquet, R. (2014b), "The Role of Energy Technologies in Long Run Economic Growth", *International Association for Energy Economics (IAEE) 37th Annual Conference*, New Yorker Hotel, New York, 16 June.

Gillingham, K., R. Newell and K. Palmer (2009), "Energy Efficiency Economics and Policy", *Annual Review of Resource Economics*, Annual Reviews, Vol. 1, No. 1, pp. 597-620, www.nber.org/papers/w15031.

Ghersi, F. and J. C. Hourcade (2006), "Macroeconomic consistency issues in E3 modeling: The continued fable of the elephant and the rabbit", *The Energy Journal*, in special issue "Hybrid modeling of energy environment policies: reconciling bottom-up and top-down", IAEE (International Association of Energy Economics), Cleveland, pp. 39-62, www.centre-cired.fr/IMG/pdf/Epreuve_EJ_17oct061.pdf (accessed 23 June 2014).

Greening, L., D. Greene and C. Difiglio (2000), "Energy efficiency and consumption – the rebound effect – a survey", *Energy Policy*, Vol. 28, Issues 6-7, June 2000, pp. 389-401.

© OECD/IEA, 2014.

Hayes, S., R. Young and M. Sciortino (2012), *The ACEEE 2012 International Energy Efficiency Scorecard*, ACEEE (American Council for an Energy Efficient Economy) Report No. E12A, Washington DC.

Hourcade, J. C. et al. (2006), "Hybrid Modeling: New Answers to Old Challenges", *The Energy Journal*, in special issue "Hybrid modeling of energy environment policies: reconciling bottom-up and top-down", IAEE (International Association of Energy Economics), Cleveland, pp. 1-12, http://halshs.archives-ouvertes.fr/docs/00/47/12/34/PDF/Hourcade_Jaccard_Bataille_Ghersi_2006_Hybrid_Modeling.pdf (accessed 23 June 2014).

Howland, J. et al. (2009), *Energy Efficiency: Engine of Economic Growth – A Macroeconomic Modelling Assessment*, report for ENE (Environment Northeast), ENE, Maine, www.env-ne.org/public/resources/pdf/ENE_ExecSum_EnergyEfficiencyEngineofEconomicGrowth_FINAL.pdf (accessed 23 June 2014).

IEA (International Energy Agency) (2013a), *Energy Efficiency Market Report 2013: Market Trends and Medium-Term Prospects*, OECD/IEA, Paris, www.iea.org/w/bookshop/add.aspx?id=460 (accessed 19 June 2014).

IEA (2013b), *World Energy Model Documentation: 2013 Version*, OECD/IEA, Paris, www.worldenergyoutlook.org/media/weowebsite/2013/WEM_Documentation_WEO2013.pdf (accessed 29 June 2014).

IEA (2012), *World Energy Outlook 2012*, OECD/IEA Paris, www.worldenergyoutlook.org/publications/weo-2012 (accessed 18 June 2014).

IEA (2011), *Saving Electricity in a Hurry*, Energy Efficiency Series, OECD/IEA, Paris, www.iea.org/publications/freepublications/publication/Saving_Electricity.pdf (accessed 23 June 2014).

Jansen, J.C. and A.J. Seebregts (2010), "Long-term energy services security: What is it and how can it be measured and valued?", *Energy Policy*, Vol. 38. No. 4, Elsevier Ltd., Amsterdam, pp.1654-1664, www.sciencedirect.com/science/article/pii/S0301421509001293 (accessed 23 June 2014).

Kronenberg, T., W. Kuckshinrichs and P. Hansen (2012), *Macroeconomic effects of the German Government's Building Rehabilitation Program*, MPRA (Munich Personal RePEc Archive) Paper 38815, University Library of Munich, Germany, http://mpra.ub.uni-muenchen.de/38815/1/MPRA_paper_38815.pdf (accessed 23 June 2014).

Lehr, U., C. Lutz and P. Ulrich (2013), *Gesamtwirtschaftliche Effekte energie- und klimapolitischer Maßnahmen der Jahre 1995 bis 2012 (National economic effects of energy-political and climate-political measures of 1995 till 2012)*, report for the Bundesministerium für Umwelt, Naturschutz und Reaktorsicherheit, Osnabrueck, GWS (Institute of Economic Structures Research) mbH, Osnabrueck, www.erneuerbare-energien.de/fileadmin/Daten_EE/Dokumente__PDFs_/gesamtw_effekte_effizienz_bf.pdf (accessed 23 June 2014).

Lehr, U., C. Lutz and D. Edler (2012), "Green jobs? Economic impacts of renewable energy in Germany", *Energy Policy*, Vol. 47, Elsevier B.V., Amsterdam, pp. 358–364, www.sciencedirect.com/science/article/pii/S0301421512003928 (accessed 23 June 2014).

Maxwell, D. et al. (2011), *Addressing the Rebound Effect*, final report for the EC DGE (European Commission Directorate-General for Environment), EC, Brussels, http://ec.europa.eu/environment/eussd/pdf/rebound_effect_report.pdf (accessed 23 June 2014).

OECD (Organisation for Economic Co-operation and Development) (2013), The OECD Env-Linkages Modelling Framework – Projecting economy-environment interactions in the coming decades, OECD, Paris, www.oecd.org/env/indicators-modelling-outlooks/flyer%20ENV-Linkages%20model%20-%20version%2025%20Sept%202013.pdf

Pollin, R. et al. (2009), *The Economic Benefits of Investing in Clean Energy: How the Economic Stimulus Program and New Legislation Can Boost U.S. Economic Growth and Employment*, report for the Centre for American Progress, Department of Economics and PERI (Political Economy Research Institute), University of Massachusetts, Amherst, www.americanprogress.org/wp-content/uploads/issues/2009/06/pdf/peri_report.pdf (accessed 23 June 2014).

Pollitt, H. (2014), Personal communication.

Popp, D., R. Newell, and A. Jaffe (2009), "Energy, the environment, and technological change", in *Handbook of Economics of Technical Change*, ed. B. Hall and N. Rosenberg, Oxford: North-Holland.

Saunders, H.D. (2013), "Historical evidence for energy efficiency rebound in 30 US sectors and a toolkit for rebound analysts", *Technological Forecasting and Social Change*, Vol. 80, No. 7, Elsevier B.V.,

© OECD/IEA, 2014.

Amsterdam, pp. 1317-1330, www.sciencedirect.com/science/article/pii/S0040162512003228 (accessed 23 June 2014).

Scheepers, M. et al. (2007), *EU Standards for Energy Security of Supply*, ECN/Clingendael International Energy Programme, Petten and the Hague, www.ecn.nl/docs/library/report/2006/c06039.pdf (accessed 23 June 2014).

Scheer, J. and B. Motherway (2011), *Economic Analysis of Residential and Small-business Energy Efficiency Improvements*, Sustainable Energy Authority of Ireland, Dublin.

Sorrell, S. (2009), "Jevons' Paradox revisited: the evidence for backfire from improved energy efficiency", *Energy Policy*, Vol. 37, No. 4, Elsevier Ltd., Amsterdam, pp. 1456-1469, www.sciencedirect.com/science/article/pii/S0301421508007428 (accessed 23 June 2014).

Sorrell, S. (ed.) (2007), *The Rebound Effect: an Assessment of the Evidence for Economy-wide Energy Savings from Improved Energy Efficiency*, report by the Sussex Energy Group for the Technology and Policy Assessment function of the UKERC (UK Energy Research Centre), London, www.ukerc.ac.uk/support/ReboundEffect (accessed 18 June 2014).

Sovacool, B. and M.A. Brown (2010), "Competing dimensions of energy security: An international perspective", *Annual Review of Environment and Resources*, Vol. 35, No. 1, Annual Reviews, Palo Alto, pp.77-108, www.annualreviews.org/doi/abs/10.1146/annurev-environ-042509-143035 (accessed 23 June 2014).

Statistisches Bundesamt (Government Statistics Office) (2013), *Aussenhandelstabelle (Foreign Trade Table)*, Statistisches Bundesamt (Government Statistics Office), Wiesbaden, www.destatis.de/DE/ZahlenFakten/GesamtwirtschaftUmwelt/Aussenhandel/Handelswaren/Tabellen/EinfuhrAusfuhrGueterabteilungen.html (accessed 23 June 2014).

Toman M. and B. Jemelkova (2003), "Energy and Economic Development: An Assessment of the State of Knowledge," *The Energy Journal*, IAEE (International Association for Energy Economics), No. 4, pp. 93-112.

Turner, K. (2013), "'Rebound' effects from increased energy efficiency: A time to pause and reflect", *The Energy Journal*, Vol. 34, No. 4, IAEE (International Association for Energy Economics), Cleveland, www.iaee.org/en/publications/ejarticle.aspx?id=2523 (accessed 20 June 2014).

UK HMRC (United Kingdom, Her Majesty's Revenue and Customs) (2013), *HMRC's CGE Model Documentation*, United Kingdom government, London, December 2013, https://www.gov.uk/government/uploads/system/uploads/attachment_data/file/263652/CGE_model_doc_131204_new.pdf (accessed 3 July 2014).

Urge-Vorsatz, D. et al. (2010), *Employment Impacts of a Large-Scale Deep Building Energy Retrofit Programme in Hungary*, report prepared by 3CSEP (Center for Climate Change and Sustainable Energy Policy) for the ECF (European Climate Foundation), ECF, Brussels, http://3csep.ceu.hu/sites/default/files/field_attachment/project/node-6234/employment-impactsofenergyefficiencyretrofits.pdf (accessed 23 June 2014).

US EPA (United States Environmental Protection Agency) (2011), "Chapter 5: Assessing the economic benefits of clean energy initiatives", *Assessing the Multiple Benefits of Clean Energy: A Resource for States*, US EPA, Washington DC, http://epa.gov/statelocalclimate/documents/pdf/epa_assessing_benefits.pdf (accessed 23 June 2014).

Wünsch, M. et al. (2014), *Benefits of Energy Efficiency on the German Power Sector*, Agora Energiewende, Berlin, www.agora-energiewende.org/fileadmin/downloads/publikationen/Studien/Energieeffizienz/Agora_ECF_RAP_System_Benefit_Study_short_version_web.pdf (accessed 23 June 2014).

Van den Bergh, J.C.J.M. (2011), "Energy conservation more effective with rebound policy", *Environmental and Resource Economics*, Vol. 48, No. 1, Springer Netherlands, Housten, pp. 43–58, http://dx.doi.org/10.1007/s10640-010-9396-z (accessed 23 June 2014).

© OECD/IEA, 2014.

Chapter 3

Public budget impacts of energy efficiency

Key points

- *By providing a more rigorous means of quantifying and monetising benefits that affect public expenditures and revenues, the multiple benefits approach corrects the misperception that energy efficiency programmes fall exclusively on the cost side of public budgets.*
- *Investment in energy efficiency holds potential to deliver additional tax revenue, provide higher returns on investment, and lower the costs of unemployment and social welfare programmes.*
- *Application of energy efficiency within the public sector itself delivers substantial cost savings through lower energy consumption, by expanding markets for energy efficient goods and services, and by reducing the fiscal drain of energy subsidies. Along with increased tax revenues from greater spending by the general public, lower public health spending and reduced investment in energy infrastructure, these effects can offset any lost revenues from energy excise duty and carbon taxes.*
- *A comprehensive benefit-cost analysis (BCA) for public budgets requires separate estimations of the cost reductions and changes to revenues; several existing modelling methods can be used to integrate these two elements into a final tally.*
- *At present, data are lacking for the positive impacts of energy efficiency in the public sector; as a result, policy impact assessments risk being incomplete and biased against energy efficiency programmes.*
- *A recent study of macroeconomic impacts from energy efficiency renovation of public buildings in the European Union showed that an annual investment of USD 56 billion through 2020 could create 760 000 jobs each year, delivering a net annual improvement to public budgets of between USD 41 billion and USD 56 billion. When broader benefits were taken into account, the figures more than doubled (USD 91 billion to USD 174 billon).*

Introduction: Emerging evidence of public budget benefits

Governments implement policies to address clearly identified needs and when the benefits of intervention outweigh the costs to society, at least in theory. Most governments have developed methods to estimate the costs to the public budget and, in many cases, the direct financial benefits of diverse policy measures. It is rare, however, that the full range of public budget benefits, in particular second-round impacts on public revenue, are estimated routinely as part of a BCA.[1]

The public budget is usually defined as a document that outlines (at national, regional or local level) government revenue and expenditure for a given year and the year following. In

1 Although the commonly used term is "cost-benefit analysis" (CBA), the IEA prefers the term "benefit-cost analysis" (BCA) due to the fact that the ratios produced are expressed as "benefit:cost". The actual approach is the same.

© OECD/IEA, 2014.

this chapter, the phrase "public budget impacts of an energy efficiency measure" is used to in reference to the public (operating and capital) revenue and expenditure related to such a measure. This publication focuses on the impact of national-level energy efficiency measures on public (national, state, federal or municipal) budgets. The intent is to investigate the value of calculating these in- and outflows to better recognise the full impact of energy efficiency measures on public budgets, i.e. the multiple benefits.[2]

Recent studies from Germany and elsewhere (highlighted throughout this chapter) demonstrate the merit of estimating both the benefits and costs to public budgets of energy efficiency policy. Research shows that current energy efficiency programmes (such as financial incentives for energy efficient buildings) are delivering net benefits to public budgets and forecast that future programmes will do the same (Kuckshinrichs, Kronenberg and Hansen, 2013; Prognos, 2013).[3] Numerous cases highlight how the public sector, particularly municipalities, has directly reduced operational costs by implementing energy efficiency measures, often with low investment costs. Yet many policy makers have not yet adopted the practice of considering the potential impact on their public budgets, for a variety of reasons (Box 3.1).

Box 3.1 Obstacles to including public budget impacts in BCA

There are diverse reasons why public budget impacts are not routinely estimated in government BCA, including:

- Energy agencies often focus on operational outputs rather than broader policy outcomes.*
- Broader public budget outcomes are perceived to be difficult to measure; government energy and environment departments often lack access to the tools to calculate these impacts.
- The scale of changes to public budgets is perceived as too small for government finance departments to prioritise in budget estimations.
- Policy makers tend to underestimate both the effort required to conduct evaluations and the value of their outcomes.
- The relationship between public and private investments and the respective returns are not clearly identified; the policies are assumed to be simple public service expenditures.
- Tax rates, and thus changes in revenue, are short-term variables; energy efficiency assessments, however, need to be sufficiently long term to capture the lifetime benefits (~20 years). This leads to a mismatch that is difficult to reconcile in analysis.
- Changes to revenue and social welfare benefits resulting from energy efficiency measures may be the responsibility of other government ministries with little interest in appraising energy efficiency policy.

* Outputs are direct results of a policy or measure, for example the number of buildings insulated in a retrofit programme, while outcomes are the changes, benefits or other effects, as result of the measures, e.g. in the buildings example this could be numbers lifted from fuel poverty or improved health.

Policy makers tend to think of energy efficiency policies in terms of their costs to the public budget. These may include the costs of: implementing regulations or standards, including information and enforcement measures; providing financial incentives such as grants, tax relief or preferential rate loans, which includes the value of the subsidy; and administering the policy.

2 In other literature, these impacts have been variously labelled "co-benefits", "ancillary benefits" and "non-energy benefits" (NEBs) – and are often used interchangeably with "multiple benefits". The International Energy Agency (IEA) uses the term multiple benefits, which is broad enough to reflect the heterogeneous nature of outcomes and to avoid pre-emptive prioritisation of various benefits; different benefits will be of interest to different stakeholders.

3 However, they do not take the energy tax revenue loss into account in the estimation.

© OECD/IEA, 2014.

Implementing a new energy efficiency policy measure or programme does incur real costs to the public budget, but these must be weighed carefully against the merits of the intervention.[4]

Classic benefit-cost assessments often overlook other, more positive impacts on the public budget. These benefits are wide-ranging and can be a mix of direct impacts (such as reduced operational and capital expenditure for the public sector) and indirect impacts (those attributed to structural changes to the economy and changed economic activity due to investment in energy efficiency). In countries with substantial levels of energy taxes, energy efficiency programmes may trigger lower energy tax revenues, which must be compensated elsewhere.

This chapter sets out to describe the range of public budget impacts from energy efficiency measures and to identify methods that can be used to estimate their value. While public budget impacts are strongly linked to wider macroeconomic impacts (highlighted in Chapter 2), there is a case for considering public budget impacts as an independent area in the appraisal of energy efficiency policy options.

The range of public budget impacts

Energy efficiency measures applied across any sector can have an impact on the public budget. While national policies and budgets are the main focus, the analysis and methods described here can equally be applied to more local energy efficiency measures and their impacts on a municipal or other sub-national budget. The range of public budget impacts from energy efficiency can be categorised into two main groups: impacts arising from the investment in energy efficiency and impacts arising from the energy demand reduction and resultant cost savings that occur (Figure 3.1).[5]

Figure 3.1 Public budget impacts of energy efficiency measures

Investment effects	
Sales tax revenue from sales of energy efficiency products and services	↑
Sales tax revenue from other goods when crowded out by Energy Efficiency	↓
Initial costs of public investment in energy efficiency products and services	↑
Social welfare and unemployment benefits expenditures	↓
Real estate transaction revenues if properties become more valuable	↓

Energy demand reduction effects	
Public expenditure on public sector energy	↓
Energy subsidies to final consumers	↓
Energy excise duty, emissions trading, and carbon tax revenues	↓
Sales and income tax revenues from sales of goods and services	↑
Public health or social welfare expenditure	↓
Public investment in energy supply infrastructure and subsidies	↓

Source: Unless otherwise noted, all material in figures and tables in this chapter derives from IEA data and analysis.

Energy efficiency investment impacts

Energy efficiency policies and/or measures often require investment in new technologies or renovation of old equipment and appliances to improve the performance of the

4 In some cases, particularly in North America, the costs of energy provider-led energy efficiency programmes are born by consumers as part of their utility bills.

5 Similar to typology set out for macroeconomic impacts in Chapter 2.

© OECD/IEA, 2014.

energy-using technology. Sometimes, more energy efficient products may not be required but rather other resources are needed (e.g. staff to implement energy management programmes in industry). In all cases, implications for the public budget are likely as investments are taxable and trickle through to create new employment in sectors both related and non-related to energy efficiency.[6]

Changes to tax revenue from sales of products and services

The market for energy efficiency goods and services has grown recently, and is expected to increase over the coming decades. With increased investment in energy efficiency goods and services, there may be changes to the tax revenues from the sales of these goods and services, but also from goods and services in other sectors. Increasingly, countries levy VAT rather than sales taxes on consumption. Globally, value-added tax (VAT) is estimated to make up 20% of government tax revenue (James, 2011). In many countries VAT or sales taxes account for a high share of government tax revenue; e.g. in France, it is 47% (Ministère de l'économie et des finances, 2013) and in China, around 33% (Hoffman, 2009).

Two related effects should be taken into account. First, an increase in sales and VAT revenues from energy efficiency goods and services may flow to governments. In most countries, VAT and sales taxes are levied at rates between 10% and 25% of the value of the good or service. Thus, if the market for global energy efficiency goods and services amounted to USD 300 billion (as it did in 2011), government revenue in this area could be between USD 30 billion and USD 75 billion.

A second effect is the impact on sales taxes and VAT on goods and services in other sectors. This may drop, if sales of other goods and services decrease as a result of investment in energy efficiency crowding out investment in other goods and services. If, however, investment in energy efficiency and the consequent reduced energy costs to consumers and businesses leads to increased economic activity overall, then sales of goods and services across a wide range of sectors may increase with resulting positive impacts on government sales tax and VAT revenues. At present, the impact on revenue from the sales of other goods and services is ambiguous and requires thorough analysis.

Box 3.2 Estimating how energy efficiency measures affect revenue from sales taxes

A recent calculation of the impact of the Kreditanstalt für Wiederaufbau (KfW) energy efficiency schemes covers two main areas: sales taxes and taxes on products, net of product subsidies (Kuckshinrichs, Kronenberg and Hansen, 2013). The study estimates that the KfW finance programme for energy efficient refurbishments and new constructions cost USD 1.3 billion (EUR 952 million) in grants and in reduced interest charged for loans. The same programme promoted gross investment in energy efficiency of USD 25.25 billion in 2011 and is estimated to have induced gross investment of a further USD 12 billion in energy efficiency in buildings, which generated tax revenues from sales incurred by investors and on products of USD 7.7 billion.

No analysis was made of the crowding out effects and the loss of revenue in energy excise duty that may have occurred, or of what investment would have taken place without the programme. Therefore, it is difficult to judge the level of additionality that has taken place. Even without this, however, the tax revenues (of USD 7.7 billion) obtained from public investment (of USD 1.3 billion) in energy efficiency relative to the programme costs appear to be very positive.

6 The IEA estimates global energy efficiency investments at USD 300 billion in 2011 (IEA, 2013).

© OECD/IEA, 2014.

Careful consideration of the additionality of any increases to sales tax and VAT revenue is needed to understand whether investment in energy efficiency causes a net change in tax revenues – especially if energy efficiency investments are subsidised. Overall spending might have remained stable if the energy efficiency investment was not made: households might otherwise have consumed other goods and services, and firms might have expanded output. If the spending on energy efficiency is not additional investment but simply shifting from other sectors, and VAT rates are consistent across all sectors, there may be no change to government tax revenues. If increasing sales of energy efficiency goods and services lead to higher profits, there may also be changes to corporate tax revenues.

Net cost of any public investment in energy efficiency

Policy measures for energy efficiency may be grouped into three broad categories: regulatory standards mandating the improvement of energy performance in a given sector; economic instruments including financial and fiscal incentives; and information measures to raise awareness and educate or provide training. From the public budget perspective, expenditures are mainly associated with the second set, reflecting the use of economic instruments to encourage investment in energy efficiency. These economic instruments often include grant schemes and concessional loans, as well as guarantees and other mechanisms to enhance credit. Policy makers are most likely to estimate the fiscal implications associated with these economic instruments. Overall, if energy efficiency policies are not cost-effective, public revenues will decline if they are implemented.

The purpose of economic instruments such as grants and concessional loans in energy efficiency policy is to provide financial incentives, through price and investment signals, to stimulate investment in energy efficiency measures. The signals may be in the form of incentives (e.g. subsidies in the form of a tax relief, grant, concessional loan or some form of direct public investment) or disincentives (e.g. price increases via taxes or the creation of a market for energy efficiency certificates). Economic instruments can also be used to leverage the amount of finance available or to improve financing terms (e.g. through reduced interest rates, the unlocking of third-party finance or dedicated credit lines). While energy pricing or taxation can raise revenue, the other economic instruments usually involve public expenditure.

Grant schemes are costly to public budgets compared to other forms of finance such as low-interest loans. From an administrative point of view, programme running costs can be lower for grants schemes than for loan schemes, as no payback needs to be administered. Good practice would involve tracking each end-use project expenditure, regardless of whether the subsidy was in the form of a grant or loan. Tax incentives and grants may subsidise the same amount but are accounted for differently in the public budget. While grants require outlay of the public budget, tax incentives impact revenues. Costs for grants may, however, be easier to track and control as they will have a direct relationship to consumers; tax incentives, by contrast, are difficult to evaluate and the amounts may only come to light at the end of the fiscal year.

Concessional loans refer to subsidies that reduce the cost of loans to investors in energy efficiency measures. There are several ways this can be done and each affects public budgets differently (Box 3.3). Public funds may be used to fund the whole loan amount through credit lines to commercial financial institutions; the funds are then on-lent at preferential rates to potential investors. Alternatively, public funds can subsidise only the interest rates and/or can provide partial debt relief for energy efficiency loans. The remaining loan amount is provided by fully participating financial institutions or third parties. The expenditure should be estimated for the amount the government has contributed rather than the full loan.

© OECD/IEA, 2014.

Box 3.3 Estimating the expenditure of a loan programme

In the case of the KfW programme for energy efficient building and refurbishment, the German government provides funds to the KfW bank to reduce the interest rates on loans for these activities. The cost of the programme comprises the interest rate reduction, the grants and a "handling margin" that KfW receives for managing the programme (Kuckshinrichs, Kronenberg and Hansen, 2013). In 2011, this amounted to USD 1.3 billion. Information is not available on the administrative cost of running the programme but it is expected to be small.

Employment impacts: Income taxes, unemployment benefits and social welfare

Studies to date show that the greatest impact of energy efficiency measures on public budgets is the reduced payout for unemployment benefits as a result of jobs created through energy efficiency programmes (Box 3.4).[7] Accurate modelling of the labour market is therefore critical to assessing public budget impacts.

Investment in energy efficiency measures, products and services can create employment in the relevant sectors or indirectly in non-energy sectors. Second-round impacts are evident when the financial gains of energy cost savings are spent on other goods (see Box 3.8 later).

Box 3.4 Estimates of employment impacts from energy efficiency measures

A recent study calculated the direct and indirect macroeconomic impacts from increasing economic activity as a result of increased investments in energy efficient renovation of buildings in the European Union (Copenhagen Economics, 2012). The results show benefits to gross domestic product (GDP) and to public finances from increased employment through higher revenues from income tax, corporate tax and VAT, coupled with reduced unemployment benefits.

The study estimates that annual investment of USD 56 billion in the energy efficient renovation of buildings through 2020 would create approximately 760 000 jobs each year. This leads to a net enduring annual improvement in public budgets of between USD 41 billion and USD 56 billion. This rises to between USD 91 billion and USD 174 billion when benefits that arise due to a one-off economic activity are included, as tax revenues rise and social expenditure to unemployment benefits drops.

The KfW energy efficiency programme was found to have induced or promoted between 121 000 and 253 300 jobs in 2011, which led to net benefits of between USD 4.1 billion and USD 13.6 billion for public budgets depending on the additional employment generated by different energy efficiency measures (Kuckshinrichs, Kronenberg and Hansen, 2013).

Other effects from investment in energy efficiency may arise, such as changes to property transaction taxes. Growing evidence suggests that investment in energy efficiency measures increases the sale value of properties (Hyland, Lyons and Lyons, 2013; Brounen and Kok, 2011; Eichholtz, Kok and Quigley, 2010).

7 This was a view strongly expressed by participants at the IEA Roundtable on Macroeconomic Effects of Energy Efficiency, Paris, January 2013.

© OECD/IEA, 2014.

Energy cost reduction impacts

A range of public budget and fiscal impacts arise for energy efficiency measures that lead to reduction in the energy costs borne by the public sector, individuals and firms. The most important are: lower public expenditure on energy consumption; benefits arising from an expanded market for energy efficient goods and services; reduced fiscal drain from energy subsidies; reduced energy excise duty and carbon tax revenues. More indirect impacts from reduced energy consumption include a spending effect that leads to increased sales of other goods and services, lower health care costs to the public budget, and reduced need for public investment in energy infrastructure.

Lower public expenditure on public sector energy consumption

The most obvious impacts on the public budget are evident when energy efficiency measures are applied directly to the public sector, generating lower energy consumption and related lower energy expenditures. This can include measures implemented in centrally and municipally owned government buildings, in water and waste utilities, in state or semi-state energy providers, in public lighting, and in institutional facilities such as schools and hospitals.

Benefits from lowered energy consumption

Although data are limited, energy use and costs in the public sector are estimated at between 2% and 5% of global energy use (World Bank, 2011). The share may be twice that high in countries with extensive district heating systems. The public sector share of heat and electricity use is even higher. For example, the public sector accounted for 9% of Brazil's total electricity in 2006 (Meyer and Johnson, 2008); in the EU-15 in 2001, the public sector was responsible for 10% of total heat and electricity consumption, with some variation among countries depending on the size of the public sector (Van Wie McGrory et al., 2002).

If the measures undertaken are cost-effective, energy demand reduction in the public sector can correspond to significant monetary benefits for public expenditure. In the European Union, the public sector owns 7% of residential buildings and 29% of non-residential buildings, so a significant share of total energy cost savings from any energy efficiency measures in buildings accrue to the public sector. Modelling results of the impacts of the directive to improve energy efficiency on public buildings show this could correspond to energy cost savings valued at USD 15 billion to USD 20 billion (EUR 11 billion to EUR 15 billion) annually in 2020, and USD 29 billion to USD 39 billion (EUR 21 billion to EUR 29 billion) in 2030 (Copenhagen Economics, 2012).[8] The public sector manages a range of infrastructure and services across varied sectors. The wide variation of public sector roles makes it impossible to estimate an average or total for the energy cost reductions possible across all activities and infrastructure. A sample of case studies illustrates the scope of public sector cost reductions that can be achieved with energy efficiency (Box 3.5).

Benefits from expanded market for energy efficient goods and services

Beyond the potential to reduce government energy bills, public procurement of energy efficiency technologies has strong potential to drive the wider market for energy efficient goods and services. Not only does the public sector account for a large share of energy use, economic activity from the public sector represents a large share of GDP and employment. In the United States, government spending at all levels (federal, state, local) accounts for 18% of GDP, while government workers (including military personnel) represent about 16% of all non-farm employment (PePS, 2014). Similarly, the public sector in the European Union

8 Commission Directive 2012/27/EU on Energy Efficiency.

© OECD/IEA, 2014.

contributes about 19% of European Union (EU) GDP. Worldwide, the percentage of GDP attributable to the public sector ranges from 10% to 25% (PePS, 2014). With these shares, the public sector can exert a major influence over the entire market, both through its direct actions and through its openly stated policies, specifications and purchasing criteria.

Box 3.5 Reducing public sector energy costs through energy efficiency

Public buildings energy management: The city of Lviv (Ukraine) launched, in 2006, a monitoring and targeting programme for energy including natural gas, district heating, electricity and water consumption in 530 public buildings. Targets for monthly utility consumption are determined annually, based on historical consumption (with the possibility to negotiate an adjustment in cases of foreseeable change in consumption patterns). Utility use is reported monthly and reviewed against the target, deviations spotted are acted upon immediately. An interesting feature is that the performance of buildings is communicated to the public through a display campaign.

This programme reduced annual energy consumption in Lviv public buildings by about 10% and tap water consumption by about 12%, translating to estimated net savings of USD 1.2 million (UAH 9.5 million) as of 2010. These significant savings have been achieved with minimal investment and recurring programme costs. A crucial initial condition for the programme was that most of the city's public buildings were already metered for energy and water consumption. Also, the city had been collaborating with international aid programmes for municipal energy since the late 1990s (EECI and ESMAP, 2014).

Retrofit of public buildings: The Federal Buildings Initiative (FBI) is a voluntary programme that facilitates energy efficiency retrofit projects in buildings owned or managed by the Government of Canada. To date, more than 80 retrofit projects have been implemented, attracting USD 312 million in private sector investments and generating over USD 43 million in annual energy cost savings. These FBI projects have demonstrated average energy cost savings of 15% to 20% and have reduced the impact of operations on the environment – cutting greenhouse gas emissions by 235 kilotonnes (kt). Other levels of government, institutions and private firms have drawn on the FBI experience for help in designing their own energy efficiency programmes (www.nrcan.gc.ca/energy/efficiency/communities-infrastructure/buildings/federal/4481).

Public procurement: The *ÖkoKauf Wien* (EcoBuy Vienna) programme in Vienna (Austria) has developed, since 2000, an internationally recognised model for sustainable municipal procurement. More than 100 ecological criteria, including energy efficiency, are used for purchasing goods and services in 23 categories, including paper, vehicles, lighting, building services, office supplies, cleaning agents, textiles and many others. Through this programme, the city has achieved annual cost savings of about USD 23.8 million (EUR 17 million) and reduced CO_2 emissions by about 30 kt. By the end of 2010, this translated into total cost savings of USD 285.6 million (EUR 204 million) and CO_2 emissions reductions of 360 kt (World Bank, 2014).

Water utility energy conservation: In Washington County, Maryland (United States) the wastewater facility carried out an equipment upgrade, involving mainly a multiple hearth furnace that incinerates biosolids. The investment cost was USD 4.5 million in 2008 but the consumption of natural gas was reduced by 76%, saving USD 400 000 per year – giving a payback of approximately 11 years. This payback period does not, however, include the benefit of avoided costs associated with delaying construction of additional incineration capacity, a benefit provided by the fact that modifications to the existing multiple hearth furnace increased capacity. Another benefit arising from the increased capacity is lower emergency haulage of un-incinerated sludge, delivering further annual savings of USD 100 000 to USD 200 000. In total, the benefits reduced the estimated payback period to 7.5 to 9 years (US EPA, 2010).

© OECD/IEA, 2014.

While governments may be concerned initially with the perceived higher up-front costs, energy efficient products have been found to decrease overall costs for public organisations by around 1% due to their lower operating costs. This value-for-money characteristic should be clearly communicated to citizens and stakeholders. Moreover, carbon dioxide (CO_2) emissions are reduced by 25% on average under green public procurement programmes (PwC, Significant and Ecofys, 2009). With lower borrowing costs than the private sector in most countries, the public sector is also well-placed to invest in energy efficiency measures.

Lower fiscal drain from energy subsidies to final consumers

Many governments pay out subsidies for both energy production and consumption. Reducing demand and supply through energy efficiency measures can significantly lower the subsidy burden within the public budget, provided the costs of the efficiency measures are less than the value of the savings obtained. The IEA estimates that fossil fuel consumption subsidies amounted to USD 544 billion in 2012, slightly up from 2011, as moderately higher international prices and increased consumption offset some notable progress that is being made to rein in subsidies. Subsidies to oil products represented over half of the amount. Even a small reduction in energy use can yield substantial ongoing reductions in the amount of subsidies paid out from public budgets.

Many of the countries providing fossil fuel subsidies are developing countries and oil exporters. The range of the subsidy rate is very large, reaching as high as 87% of the full cost of the fuel supply in some countries. In the Organisation for Economic Co-operation and Development (OECD), some member countries target fuel and energy subsidies to low-income groups to compensate for high heating bills in winter. Investment in energy efficiency measures in residential buildings can improve the quality of the buildings, lowering energy bills and negating the need for subsidies to help pay those bills. Reducing the amount of energy consumed through energy efficiency measures can have a significant impact on the public expenditure on these subsidies; it does, however, raise the question of who should pay for the energy efficiency improvements if the buildings are not publicly owned.

Less data are available on the value of subsidies for energy production across different countries (Box 3.6). In the European Union, most energy production subsidies aim to either support early retirement of coal-fired power plants that are not affected by energy efficiency measures and/or stimulate renewable energy production. By implementing energy efficiency measures and reducing demand for energy, it becomes easier to meet renewable energy targets and reduce the subsidy provided for renewable energy. Copenhagen Economics estimates that renovation of the EU building stock will lead to an energy demand reduction of 5.4% to 8.9% and can reduce subsidy outlays to renewable energy deployment by USD 9.6 billion (EUR 7.1 billion) annually in 2020, constituting a significant improvement to public budgets. Other issues must be considered, however; for example, if renewable capacity replaces existing fossil fuel-based capacity, it could lead to higher levels of stranded assets and increased demand from power companies for government support or compensation.

© OECD/IEA, 2014.

Box 3.6 Addressing the public costs of energy subsidies

The *OECD Inventory* (OECD, 2013) demonstrates that some member countries are paying out significant energy-producer subsidies; reducing energy demand through energy efficiency measures could help substantially lower these costs.

Germany: Production of hard coal in Germany has traditionally attracted government support for geological, historical and political reasons. The total nominal value of estimated producer support for hard coal amounted to about USD 6.8 billion (EUR 5 billion) (0.3% of GDP) in 1999. As domestic production of hard coal remains largely uneconomic, the government has decided to phase out its support to the industry by 2018. This gradual phase-out reduced the total amount of estimated producer support by more than half, or about USD 2.7 billion (EUR 2 billion) (0.1% of GDP), in 2011.

Poland: The bulk of state aid to the energy sector in Poland is apportioned to the coal industry. The total amount of producer support for coal over the 1999-2011 period is estimated to have exceeded USD 7 billion (PLN 25 billion). Prior to the collapse of communism, coal production was mainly supported through the regulation of coal prices and the provision of various social benefits to coal miners. With the economic transition of the 1990s, the state began to restructure the coal sector through a series of capacity-adjustment programmes, which resulted in closure of unprofitable mines and reduced employment. These programmes proved ineffective, however, and the state decided (as in Germany) to gradually phase out government support. Most remaining costs are now associated with historical liabilities. Poland, similar to other coal-producing member states of the European Union, is subject to European Council regulations regarding state aid. Since 2011, Council Decision 2010/787/EC authorises state aid only for the purpose of closing mines, the treatment of health damages sustained by miners, and addressing the environmental liabilities related to past mining.

United States: In the case of the United States, the *OECD Inventory* estimates that total producer support, including tax expenditures at the federal level and for some states, represented about USD 6 billion in 2011 (about 0.04% of GDP). The federal budget for fiscal year 2013 proposes to eliminate a number of tax preferences benefiting fossil fuels, which could increase revenues by more than USD 23 billion over the years 2013 to 2017 (OMB, 2012). Some measures can also be found at the sub-national level, where states sometimes provide additional tax expenditures benefiting oil and gas producers. Based on a sample of ten coal- or oil-producing states (Alaska, California, Colorado, Kentucky, Louisiana, Oklahoma, Pennsylvania, Texas, West Virginia and Wyoming), the OECD found that sub-national measures accounted for about 53% (USD 3.1 billion) of the USD 5.8 billion total estimated producer support in 2011.

Source: OECD (2013), *An OECD-Wide Inventory of Support to Fossil Fuel Use and Production*, OECD Publishing, Paris.

Reductions in energy excise duty and carbon tax revenues

Reduced energy demand is a key objective of energy efficiency measures. However, as most countries levy taxes or excise duty on energy, and in some cases the associated CO_2 emissions (unless there is a subsidy or an exemption), lower energy consumption will lead to lost government revenue. Energy tax revenue is a significant source of revenue for governments: in the OECD, it makes up between 1% and 5% of GDP (OECD, 2014). Revenue from transport fuel excise duty in particular is substantial, amounting to nearly 2% of GDP in some European countries (Box 3.7). In the EU-15, for example, transport fuel taxes amounted to nearly EUR 200 billion in 2012, representing important sources of revenue for the public budget (ACEA, 2013). Fuel taxes are an efficient means to allocate transport system development and maintenance costs to system users.

In order to maintain a stable level of revenue, governments may need to increase fuel charges as vehicle fuel economy improves. Such a move may be difficult to communicate to stakeholders, as consumers will not appear to reap the benefit of their investment in more

© OECD/IEA, 2014.

fuel-efficient vehicles. Nonetheless, the energy prices are lower to reach the same policy targets than would have been needed had cost-effective energy efficiency improvements not been undertaken.

Box 3.7 Changes in revenue from energy excise duty

Modelling the macroeconomic impacts of residential building renovation in the European Union suggests that the expected reduction in energy consumption would give rise, in different scenarios, to tax revenue losses between USD 7 billion and USD 9.8 billion annually in 2020. If the energy efficiency potential for 2030 is met, energy taxes would be reduced annually by a total of between USD 13 billion to USD 18.7 billion in 2030 (depending on the scenario). Any loss of tax revenue should not be considered as a loss to society as a whole, since it is a transfer from governments to consumers (i.e. the money remains within society), but it still counts as a cost to public expenditures (Copenhagen Economics, 2012).

In a 2012 macroeconomic modelling and tax revenue impact study for four Canadian provinces, Environment Northeast (ENE) used a representative sample to conduct a high-level assessment of energy sales tax losses from reduced fuel sales and the net tax revenue impact. The study found that the average annual direct sales tax lost was equivalent to USD 49 million (CAD 52.7 million in 2011 CAD) at the provincial level and USD 87 million (CAD 93.3 million) for the federal government.

Sales of goods and services due to spending effect from energy cost savings

Energy cost-saving measures targeted at both household and company level can lead to increased disposable income. This augmented income can be used in a number of ways (as outlined in Chapter 2 on macroeconomic impacts), leading to indirect impacts and economic activity that may or may not impact the public budget. The energy savings may be used in one of three ways: they may be saved; they may be reinvested in other energy efficiency goods and services (paying off the initial investment or acquiring new goods and services); or they may be spent on other, unrelated activities that may or may not consume energy. The public budget is likely to be impacted through changes in revenues from sales and import taxes, income taxes related to changes in the numbers employed, and any associated social welfare costs or benefits. Depending on the level of energy cost savings achieved, this impact may be the largest and most durable, as the savings from the energy efficiency measures continue over a long period.

Comparing the impact of these mainly positive effects with the drop in fuel taxes provides interesting insights. In most countries, fuels are taxed at a higher rate than other goods (excise duties) so shifting consumption from fuel to other goods will lead to lower tax revenues. But in the United Kingdom, the reduced VAT rate for heating fuels means that tax revenues would actually go up if people spent less on heating and more on household goods. As demonstrated, the net effect is country-specific.

Box 3.8 Net effects of changes to taxes on other goods and services

The lifetime savings resulting from vehicle fuel economy standards in California are estimated to range between USD 5 000 and USD 7 000 for each vehicle/owner (Roland-Holst, 2011). In this scenario, about 70% of household spending and a significant portion of enterprise spending on non-energy inputs shift to services from fossil fuels. The study examined the wide variation in

© OECD/IEA, 2014.

labour intensity of different economic sectors and concluded that the resulting expenditure shifting leads to substantial net job creation. It also pointed out that the additional demand for goods and services is not necessarily in "green" sectors but is more likely to be in services sectors. Thus, many of the jobs created are in the services sectors, which are the most labour intensive. Ultimately, this can lead to substantial revenue raised through income taxes and the VAT on goods and services.

Indirect impacts from energy demand reduction and increased consumer spending are shown to lead to job creation and gross national product (GNP) growth, in turn delivering revenue income for the public budget through VAT, sales taxes, income tax and social charges (Prognos, 2013). In the study's baseline scenario, the increase in tax revenue was estimated at USD 8.2 billion (EUR 6 billion) while the "optimistic" scenario saw a rise to USD 36.6 billion (EUR 27 billion).

Copenhagen Economics estimate the value of energy cost savings from buildings renovation in Europe 2012-20 to be USD 89.5 billion to USD 127.5 billion (EUR 66 billion to EUR 94 billion) per year. If this is spent on goods and services, the resulting impact of public budgets should be very substantial.

In Canada, a tax revenue impacts assessment indicates that the increase in economic activity generated by the energy efficiency investment and savings would lead to an average annual net increase in corporate, personal, and sales tax revenue of USD 259 million (CAD 281 million) at the provincial level and USD 287 million (CAD 312 million) for the federal government (ENE, 2012). While more detailed analysis is warranted to determine the actual magnitude of the impacts, the direction of the impact is clear. Energy efficiency measures will reduce sales tax collections from the sale of energy; however, the loss will be more than offset by increased tax collection resulting from the efficiency-driven increase in economic output.

General household spending in Germany is predicted to rise by 0.08% or USD 14.4 billion by 2020 (compared with a reference scenario) as a result of the energy efficiency measures planned. This is mainly due to increased demand for energy efficient goods and services, and a second-round impact in which energy bill savings (the consumer surplus) are spent on other goods (Lehr, Lutz and Pehnt, 2012).

Impact on public health budgets

The link between energy efficiency measures that improve the quality of indoor and outdoor environments and the health impacts such measures generate is a new field of investigation. The discussion here on related implications for public budgets is one element of a wider investigation covered in Chapter 4. A growing body of evidence supports the claims that energy efficiency measures – in residential and commercial buildings, transport and industry – have positive impacts on public health.

Studies to date have focused primarily on the public health impacts arising from the improved quality of indoor environments and from reduced emissions from transport and energy generation.

- In buildings, measures to improve insulation, heating and ventilation systems can have positive impacts in reducing respiratory and cardiovascular diseases, allergies, arthritis and rheumatism. They also drive significant and consistent mental health improvements (Liddell and Morris, 2010).
- Improved efficiency in transport and power generation systems resulting in reduced local air pollution has demonstrated potential to lower the incidence of respiratory disease, as well as conditions related to inadequate physical activity and traffic-related stress and injuries. Other measures across all sectors that lead to reduced air pollution will have significant health benefits (WHO, 2011).

© OECD/IEA, 2014

Several studies show that the health and well-being impacts alone could actually outweigh the economic benefits of energy demand and emission reductions, in terms of both financial and social value (see Chapter 4). Financial benefits are reflected in avoided hospitalisation and pharmaceutical costs, as well as the cost savings from fewer days absent from work or school, all of which can lead to reduced public health budget expenditure (Box 3.9). As lower income groups are likely to have disproportionately more of their health costs covered by the public budget than other income groups, energy efficiency measures that have health benefits for these groups should be prioritised.

Box 3.9 Estimations of public health benefits

BCA carried out in the area of health and well-being benefits of energy efficiency show strong evidence of multiple benefits (Chapter 4), not only in terms of avoided costs for the public budget but in relation to productivity and education. The most significant results were found among low-income households where pre-existing health problems had been identified. *Warm Up New Zealand: Heat Smart,* a three-year energy efficiency retrofit programme implemented by the government, was estimated to deliver an overall benefit-cost ratio of 4:1, with 99% of this impact being made up of health benefits, mostly arising from the installation of energy efficient insulation. (The programme is described in more detail in Chapter 4, and the model used in its evaluation is described in Box 2 in the Companion Guide at the end of this publication).

The ExternE project set out to quantify the health benefits associated with reducing air pollution from energy generation in EU member states, including the development of *EcoSense,* a tool to carry out the calculation (ExternE, 2011). Additional estimates conducted by Copenhagen Economics focus on the health benefits of reduced air pollution due to less energy production and improved comfort through buildings renovation. The study finds a high rate of uncertainty in values and therefore estimates the public health budget benefits for EU member states to be between USD 16.9 billion and USD 60.3 billion (EUR 12.4 billion and EUR 44.3 billion) per year. This is a subset of the total health benefits of USD 57.2 billion to USD 119.8 billion (EUR 42 billion and EUR 88 billion) per year under low- and high energy efficiency scenarios, respectively. The two scenarios see primary energy demand reduced by 5.4% (low energy efficiency scenario) and 8.2% (high energy efficiency scenario) requiring investment of between USD 55.8 billion and USD 106.2 billion per year to 2020, respectively.

Reduction in public investment in energy supply infrastructure

Energy prices comprise both fixed and variable costs. The fixed costs (e.g. expenditures for labour, maintenance and debt repayment related to the initial investment) are determined by ongoing costs associated with operating the energy infrastructure and do not vary with output volumes. Reduced energy demand due to efficiency measures can be important in reducing the need for investments in additional energy generation capacity and the operation and maintenance of energy infrastructure. Since in many countries energy infrastructure is owned (or at least part-owned) by the state, this leads to further savings for the public budget.

Countries with older energy infrastructure face a period of heavy investment. Ecofys (2013) cites the example of Poland, where 37% of installed electricity capacity is 30 to 40 years old. To meet currently projected demand, the estimated investment needed for the Polish electricity system by 2030 is USD 122 billion to USD 136 billion. The National Energy Conservation Agency calculates that Poland has the potential to save 30% to 35% of current energy demand through energy efficiency measures. These data illustrate the potential of energy efficiency to avoid energy infrastructure investment (Box 3.10).

© OECD/IEA, 2014.

Box 3.10 **Reducing future infrastructure costs through energy efficiency measures**

Energy efficiency measures can offset future rising demand for electricity.* The IEA *World Energy Outlook 2012* estimates in its Efficient World Scenario that such measures could reduce by 16% (against a baseline scenario) the required investments in generation capacity as well as transmission and distribution infrastructure between 2012 and 2035, bringing the investment cost down to USD 5.9 trillion. These savings are estimated to offset half of the additional investment needs on the demand side.

Modelling by the European Climate Foundation (ECF, 2011) estimates that the energy efficiency measures needed to achieve the EU target of 20% improvement in energy efficiency by 2020 can reduce electricity demand growth by 83% compared with the decarbonisation scenario (On Track scenario), in which electricity is decarbonised and demand from renewable sources is increased. This results in a 50% decrease in transmission investment and a 31% decrease in back-up investment, translating to a reduction of USD 408 billion (EUR 299 billion) in investments (or 30% lower capital expenditure).

* In many countries, decarbonisation of the energy sector implies increasing electrification of the transport and heating sectors, causing increased pressure on electricity infrastructure.

Also of relevance in this area is a trend of governments seeking to minimise their asset holdings. This does not address ownership; rather it focuses on reducing operational energy costs and the impact on public budgets of strategically using energy efficiency policies to avoid future investments.

Methodological approaches

This section sets out recent experience with estimating the public budget impacts of energy efficiency measures, and outlines some options for policy makers to consider in undertaking such an assessment as part of a policy appraisal process. As outlined in the Companion Guide at the end of this publication, policy appraisal includes an assessment of the range of public and private costs and likely impacts of a policy. If possible, these impacts are quantified and monetised as part of a BCA. This allows policy makers and the public, once results are published, to understand the rationale and consequences for implementing the policy. This section examines how to measure the public budget impacts described in the sections above.

Estimating public sector impacts

To date, there is relatively little experience with estimating the full public budget impacts of energy efficiency measures and programmes.[9] While the public budget costs are generally calculated for implementing new energy efficiency policies and programmes, the full revenue implications, both positive and negative, do not appear to be routinely estimated as part of the policy appraisal process.

In the wider literature, a few studies have explicitly set out to examine the public budget aspects of energy efficiency policies (Kronenberg, Kuckshinrichs and Hansen, 2012; Prognos, 2013; Copenhagen Economics, 2012). As seen in the previous section, the impacts can be grouped into direct and indirect impacts arising from energy demand reductions and from

9 Although precedent has been set in sectors such as renewable energy, transport or information technology.

© OECD/IEA, 2014.

investment in energy efficiency measures. The methodological options to estimate the two kinds of impacts are quite different and therefore are described separately in the next subsections. Energy cost reductions accruing to the public sector from energy efficiency measures are the most straightforward and common to estimate. Most of the other public budget impacts are fiscal effects arising through changes to tax and excise duty revenues, and through effects on public health budgets. These are more difficult to estimate and, at present, are rarely calculated as part of energy efficiency policy impact assessment or appraisal.

Estimating public sector energy cost reductions

Several well-established methodologies used to estimate energy cost savings when evaluating energy efficiency programmes can be adapted to calculate the impacts of public sector energy efficiency programmes. The following steps can be taken:

- An energy audit should be undertaken to assess current energy use in the public sector. Over time, this can be developed and extended to encompass monitoring of energy use in the public sector (SEAI, 2013).
- The areas in the public sector that are targeted and/or likely to be affected by the energy efficiency measure should be identified, counterfactual (or baseline) energy use established and market readiness analysed.
- High and low estimates of the energy cost savings that can be achieved by the policy measure should be calculated by fuel type.
- The value of the energy bill savings should be calculated, using the current and projected energy prices paid by the public sector for electricity, gas, oil and any other fuels used.[10]
- The value of avoided energy imports should be estimated.

Estimating changes in public budget revenue

Many of the public budget impacts of energy efficiency programmes are driven by changes to tax revenues. These impacts have generally been estimated as add-ons to macroeconomic assessments of energy efficiency programmes; thus, the methodology is tied directly to macroeconomic assessment of energy efficiency policies and measures. The main modelling methods are described in Chapter 2; the Companion Guide at the end of this publication also provides a short summary, focusing on the estimation needed on top of a macroeconomic assessment.

Many revenue effects are second-round (indirect) effects that occur as a result of energy efficiency measures causing significant and prolonged changes to certain economic parameters, such as the level of investment, amount of energy cost savings and spending, GDP, employment, etc. All of these factors can translate into a revenue impact that can be calculated either through a relatively straightforward basic assessment with the application of multipliers or through more complex modelling.

Basic assessment

A **basic assessment**, which can be carried out using a spreadsheet, is the most commonly used method to estimate revenue effects from energy efficiency measures. It is usually limited to the impacts of a single sector, as it is difficult to represent cross-sectoral linkages in a basic assessment. To calculate the public budget implications, the estimation

10 There are a range of possible energy prices that could be used to estimate the avoided energy cost to government: retail costs, long-run marginal costs, or average wholesale costs. This may be more complicated still if the government is involved in energy supply.

© OECD/IEA, 2014.

uses key macroeconomic values, such as public investment, GDP and sectoral economic activity, and employment – all of which should have already been estimated as part of a macroeconomic assessment of the impacts of the energy efficiency programme. Inputs required for a basic assessment include the costs of the programme, the amount and value of the energy cost savings, the investments needed, the GDP impact, the employment created or lost, and the sectors in which increased spending is likely to occur.

A set of factors, tax rates and multipliers can be applied to these input variables to carry out the calculation of public budget impacts (Table 3.1). It is important to think about the likely duration of the effects, as some may be relatively short-lived, while others may be more enduring. The individual estimates of public budget impacts should be added to the programme costs and to the direct energy cost-saving benefits to the public sector (already estimated, if applicable).

Table 3.1 Basic assessment of public budget impacts from energy efficiency programmes

Input variable	Measure	Impact
Energy saved in public sector	Energy prices	Public sector energy bill
Change in GDP	Fiscal multiplier	Public budget impacts due to economic activity
Investment in all sectors	Employment factors	Jobs created Changes to unemployment and social welfare benefits Crowding-out effects
Investment in energy efficiency goods and services	VAT rates	VAT revenue
Employment created	GVA/employee	GDP change
Employment created	Unemployment benefits, income tax rates	Income tax revenues, avoided unemployment benefits
Energy saved	Energy tax or subsidy rates Carbon tax/emissions trading scheme	Change in energy tax (excise duty) revenue or subsidy bill Change in emission tax revenues

Note: GVA = gross value added.

The simplest – but least transparent – method to calculate the change in public spending is to apply fiscal multipliers to any change in GDP.[11] Fiscal multipliers are usually published by governments; those for the EU-27 provide an example (Table 3.2). Multipliers should be applied with caution, as they can vary if the economy structure or circumstances change. Fiscal multipliers tend to be larger in crisis periods due to the higher share of financially constrained consumers. Assessing the current size of fiscal multipliers is complex because their value depends on their composition, their durability, and on the economic environment at large (Boussard, De Castro and Salto, 2012). However, multipliers can be useful for a quick estimate of the public budget implications from a change in GDP as a result of a policy.

11 Fiscal multipliers provide a ratio between government spending and the expected return in revenues that results. In other words, if the multiplier is valued at 0.4 then for every Euro of government spending, EUR 0.4 is returned in the form of revenues from increased economic activity.

© OECD/IEA, 2014.

Table 3.2 Fiscal multipliers for EU-27

Country	Fiscal multipliers (2011)
Austria	0.47
Belgium	0.51
Bulgaria	0.33
Cyprus*	0.43
Czech Republic	0.36
Denmark	0.65
Estonia	0.30
Germany	0.51
Greece	0.42
France	0.53
Finland	0.58
Hungary	0.44
Ireland	0.44
Italy	0.49
Lithuania	0.29
Luxembourg	0.44
Latvia	0.30
Malta	0.38
Netherlands	0.62
Poland	0.38
Portugal	0.45
Romania	0.32
Spain	0.43
Sweden	0.61
Slovenia	0.45
Slovakia	0.33
United Kingdom	0.46
Average EU-27	**0.44**

* 1. Footnote by Turkey: The information in this document with reference to "Cyprus" relates to the southern part of the Island. There is no single authority representing both Turkish and Greek Cypriot people on the Island. Turkey recognises the Turkish Republic of Northern Cyprus (TRNC). Until a lasting and equitable solution is found within the context of United Nations, Turkey shall preserve its position concerning the "Cyprus issue".
2. Footnote by all the European Union Member States of the OECD and the European Commission: The Republic of Cyprus is recognised by all members of the United Nations with the exception of Turkey. The information in this document relates to the area under the effective control of the Government of the Republic of Cyprus.
Source: Boussard, J., F. De Castro and M. Salto (2012), *Fiscal Multipliers and Public Debt Dynamics in Consolidations*, Economic Papers 460, Directorate-General for Economic and Financial Affairs (DG ECFIN), Brussels.

© OECD/IEA, 2014.

Box 3.11 Estimation of public budget expenditure for an EU loan programme

Copenhagen Economics used the basic assessment method to estimate the macroeconomic effects of building renovation in Europe over a short- to medium-term horizon, applying a series of cost curves and publicly available fiscal multipliers.

- The starting point in this estimation was the calculation of the energy demand reduction achievable under two scenarios of high and low energy efficiency ambition.
- From there, the investment required to achieve the desired energy demand reduction was calculated using technology cost curves.
- Based on the investment needed, the potential for employment was estimated using multipliers from a literature review.
- Next, the published GVA values per employee for the construction sector were used to calculate the impact on GDP.
- Fiscal multipliers were then applied to the GDP to calculate the change in public budget revenues.

This method provides a relatively straightforward estimate of the public budget impacts of a single sector.

Source: Copenhagen Economics (2012), *Multiple Benefits of Investing in Energy-efficient Renovation of Buildings – Impact on Finances*, report commissioned by Renovate Europe, Renovate Europe, Brussels.

More advanced modelling of tax revenue implications of energy efficiency programmes

More complex modelling approaches, which cover the whole economy and can better represent transfers among sectors, can also be useful for public budget estimation and assessment. These approaches are generally preferred to assess the macroeconomic impacts and provide a way to estimate the public budget impacts via macroeconomic models. Because of their broad coverage, most estimates of the macroeconomic impacts of energy efficiency policies are performed using input-output (I-O) analysis, computable general equilibrium (CGE) or macro-econometric models. Some analyses have developed a module to estimate the public budget impacts. Only the implications for public budget estimation are given here, but these methods are described in more detail in the Companion Guide at the end of this publication.

- **I-O analysis** uses purchases between sectors to estimate how increased spending in one sector affects spending in others. Although it is a relatively static and rigid method, and involves only one year's data, I-O is useful in estimating second-round impacts such as VAT revenues from the sales of different goods and services, as it represents the shifts in purchases among sectors. Some countries provide annual or bi-annual I-O tables so trends over time can be identified. However, the I-O structure does not adjust to represent lower energy input as a result of energy efficiency. Almost all major macro-econometric models are built around a nucleus of I-O identifiers and parameters. Recently, KfW used I-O analysis to estimate public budget impacts of an energy efficiency refurbishment and construction programme in Germany (Box 3.12).
- **CGE models** can identify subtle linkages between different economic sectors, and can be used to model the macroeconomic impacts of energy efficiency policies, leading to employment effects, which in turn allows for income tax revenue changes. The Berkeley Energy And Resources (BEAR) model has been used to assess the economic impacts of changes to Californian fuel economy standards. Although the public budget impacts are not estimated explicitly, the shift in jobs from high energy-intensity sectors to the services

© OECD/IEA, 2014

sector is analysed with a calculation of a net jobs increase. The increase in disposable income for households over the lifetime of more efficient vehicles is also estimated. If combined with an I-O calculation of the spending effect, this analysis could assess the change to sales tax and excise duty revenue.

Box 3.12 Input-output model to estimate public budget impacts in Germany

When estimating the impacts of the KfW energy efficiency refurbishment and construction programme, researchers extended an I-O model to incorporate a module that simulated effects on public budgets. This module captures all the public revenue and expenditure data relevant in the context of these KfW programmes and allocates it to the appropriate administrative authority level (federal, state or municipal) and to social insurance funds. The findings show that the amount of public budget effects depends significantly on the degree of additional manpower required. The study does not estimate changes in revenue from energy excise duty or the value of changes in energy consumption in the public sector, as those may not be relevant to the KfW programme.

- **Macro-econometric** models are economy-wide models based on estimates of historical relationships, which are assumed to continue over time. Such models use econometric analysis as described above but cover the whole economy. Structural change can be difficult to model with econometric models due to their reliance on historical data (this can be an issue for all models). In the Energy-Environment-Economy Macro-econometric model for Europe (E3ME), the I-O coefficients are adjusted to reflect reduced energy demand as a result of energy efficiency. It can also capture changes in energy- and emission-related revenues.

A challenge to modelling the public budget implications of energy efficiency measures with both CGE and econometric models is that policy measures are often assumed to be "revenue-neutral", meaning that the direct fiscal effects are made to balance (e.g. a tax rate is increased to pay for the energy efficiency). This reflects the difficulty in representing the full tax system in models, which is due to both the data and the "representative agent" assumption that all individuals are the same and react in the same way to price changes, which can lead to misrepresentation of tax revenues. In addition, some of the larger taxes (e.g. on property transactions) are difficult to fit into a modelling framework. Changes in the expenditure on social benefits, which can be quite large, are even more complex and non-linear to model.

Subsidy, entitlements and tax incidence vary greatly among countries, so modellers need to carefully construct country-specific representations to accurately predict public budget and revenue impacts. Microsimulation models are a better tool for investigating tax revenue implications. Euromod, for example, splits households into quite detailed groups with different income and consumption patterns. The disadvantage of this approach is that it assumes constant GDP. Ideally, microsimulation and macroeconomic models would be used in combination. The OECD has a tool to estimate welfare benefits that could perhaps be adapted to energy efficiency impacts.[12]

Finally, similar to other energy efficiency benefits, many public budget impacts stem from changes in energy demand. Thus, it is particularly important to address the issue of rebound effects within the modelling framework (Rebound effect perspective 3).

12 www.oecd.org/social/soc/benefitsandwagestax-benefitcalculator.htm.

© OECD/IEA, 2014.

Rebound effect perspective 3

Analysis of public budget impacts

The impacts on the public budget from energy efficiency measures can both cause and be affected by rebound effects, which reduce the real energy demand reductions compared with those expected from energy efficiency measures. The rebound effect can take two forms in relation to public budgets.

First, reduced spending on energy in the public sector can cause direct rebound effects, where for example the savings are subsequently spent on other measures in the public sector, causing an increase in energy use and negating some of the reductions in energy demand achieved through the energy efficiency measures. Additionally, higher than foreseen public spending may lead to a rise in economic activity in general with knock-on increases in energy consumption. This type of rebound effect depends on the energy cost reductions achieved for the public sector and budget implications. It is linked to an overall boosting of consumption that counteracts potential energy demand reduction.

Second, the rebound effect in other sectors can also affect some of the public budget impacts outlined in this chapter. If the rebound effect causes a lower energy demand reduction than expected across the economy or in specific sectors, the reduction in energy excise duty due to energy demand reduction is likely to be mitigated. Similarly, sales tax and VAT revenues in other sectors may also be affected if the spending effect from predicted energy cost reduction does not occur.

Other public budget impacts that result from revenue associated with increased investment in energy efficiency improvements can arise independent of energy demand reductions and are not affected by the rebound effect.

Policy-making considerations

Estimation of energy efficiency impacts on the public budget should be undertaken with care. The following cautions should be considered in conjunction with the potential pitfalls already highlighted in relation to macroeconomic impacts (Chapter 2).

Issues around estimating **employment effects** from energy efficiency programmes are particularly relevant to the public budget. The duration of any jobs created (i.e. whether they are considered long or short term) is important, as is the sector in which they are created as this determines the quality and wage level. The balance of losses and gains, as jobs shift from one sector to another, should be examined in calculating the net jobs effect. The economic structure and current level of employment will determine whether new investment will create new jobs or simply shift jobs from other sectors.

A model of two employment scenarios – "overtime" and "jobs" – is revealing (Kuckshinrichs, Kronenberg and Hansen, 2013). In the first scenario, all of the additional demand for labour is met by current employees in the construction sector working overtime. In the jobs scenario, new jobs are created and filled by individuals who were previously unemployed. The difference between the two scenarios for the public budget is USD 1 720 million in favour of the jobs scenario; when more people join employment, public expenditure on unemployment costs is avoided.

Crowding out is an issue for public budget impacts. If a public programme and financial support for energy efficiency measures crowds out private investment, crowding out of other potential tax revenues may follow. Just as it is necessary to consider whether energy efficiency-related employment and investment are additional, tax revenue should be examined for additionality.

© OECD/IEA, 2014.

Energy prices are crucial in determining the value of energy cost savings; it is important to make the right assumptions about current and future energy prices, and how they may be affected by the policy considered.

Discount factors are vital for estimations of public budgets in order to convert future costs or benefits to present values, using an annual percentage rate at which the present value of a future unit of currency is assumed to fall away over time. Most public sectors use a standard discount rate; for example, *The Green Book* used in the United Kingdom sets this rate at 3.5%.[13] When the time frame for the estimation is very long (e.g. greater than 30 years), a declining discount rate is applied (Table 3.3). If set incorrectly, discount rates can lead to significant over- or under-estimation of the public budget effects in the future.

Table 3.3 Declining long-term discount rate applied in the United Kingdom

Discount rate	3.5%	3.0%	2.5%	2.0%	1.5%	1.0%
Period of years	0-30	31-75	76-125	126-200	201-300	301*

* UK HM Treasury (2003) *The Green Book: Appraisal and Evaluation in Central Government*, United Kingdom government, London.

Some countries take a different approach to discount rates. New Zealand, France and Canada, for example, use the social opportunity cost of capital, which is the return foregone by investing in one project rather than in an alternative project. It assumes that the budget is limited, i.e. the project cost is the value of the next best alternative foregone. The discount rates used for public sector projects in New Zealand vary according to the application (Table 3.4).

Table 3.4 New Zealand discount rates

Weighted average cost of capital	Applications
8.0%	Default
6.0%	Buildings
8.0%	Infrastructure
9.5%	Technology

Source: New Zealand Treasury (2008), *Public Sector Discount Rates for Cost-Benefit Analyses*, New Zealand Treasury, Wellington.

Further research for stakeholders

Despite being a new area of evaluation in relation to energy efficiency programmes, early work in public budgets is delivering interesting results; it is clear that the topic warrants further exploration (Table 3.5). A key challenge is that more data are needed to make more robust estimations of the public budget impacts of energy efficiency.

13 *The Green Book* is the guidance document provided by UK HM Treasury (United Kingdom Ministry of Finance) to public sector bodies on how to appraise proposals before committing funds to a policy, programme or project.

© OECD/IEA, 2014.

Table 3.5 Further stakeholder research and collaboration opportunities in public budget impacts

Area	Specific actions
Benefit areas and causal linkages	New research efforts to reinforce information on macroeconomic impacts should be pursued as this will also advance understanding of public budget impacts. Research needed to adapt macroeconomic models handling energy efficiency measures to include tax revenues and other public budget implications.
Data, indicators and metrics	Ministries should begin collecting data on public sector energy consumption to support increased efforts to measure the benefits in this area. Also, better data are needed on tax revenues disaggregated by products and services so that the impact of energy efficiency measures can be estimated.
Assessment methodologies	Macroeconomic modelling, energy audits of public sector facilities and epidemiological health studies relating to impacts of energy efficiency should be pursued.
Collaboration initiatives	Ministries should work together to develop coherent record-keeping processes on tax, VAT and excise duty revenues related to energy efficiency measures across ministries so that results can be compared and compiled in the context of government-wide assessments of multiple benefits of such measures. Information relating to the public budget impacts of energy efficiency measures should be communicated to non-energy ministries.

Conclusions

At present, the full public budget impacts of energy efficiency programmes do not appear to be routinely estimated as part of policy impact assessments carried out by governments. A key reason for this may be that some of these impacts lie outside the remit of the usual government ministry (of energy) or agency administering energy efficiency programmes. Moreover, the estimation of public budget impacts is linked closely to estimation of the macroeconomic effects of energy efficiency programmes (a complicated endeavour in itself) which are used as inputs to assess the public budget impacts.

As with other areas, the public budget impacts from energy efficiency programmes can be separated into two broad categories: effects arising from investment in energy efficiency; and effects relating to reduced energy consumption and costs. These effects may touch both the public sector itself and actors in other sectors, such as households and industry.

The effects arising from increased investment in energy efficiency goods and services include changes to tax revenues for governments from sales taxes, income taxes and avoided unemployment benefits if jobs are created. There may be net costs to the public budget arising from investment in energy efficiency goods and services. These effects may be short-lived, as they may be associated with one-off investments.

The effects linked to reduced energy consumption, by contrast, are likely to be more durable and may cause direct improvements in the public budget over long periods. This is particularly true where they relate to energy efficiency programmes that reduce expenditure on energy within the public sector. While the costs to the public budget of energy efficiency programmes generally are estimated, without the estimation of all positive and negative public budget impacts – such as the reduced outlay on public sector energy and potential changes in tax revenues – policy impact assessments risk being incomplete and biased against energy efficiency programmes.

© OECD/IEA, 2014.

Bibliography

ACEA (European Automobile Manufacturers Association) (2013), *ACEA Tax Guide 2013*, ACEA, Brussels, www.acea.be/uploads/publications/ACEA_Tax_Guide_Highlight-EU_Summary_2014.pdf (accessed 21 June 2014).

Boussard, J., F. De Castro and M. Salto (2012), *Fiscal Multipliers and Public Debt Dynamics in Consolidations*, Economic Papers 460, Directorate-General for Economic and Financial Affairs (DG ECFIN), Brussels, http://ec.europa.eu/economy_finance/publications/economic_paper/2012/pdf/ecp460_en.pdf (accessed 21 June 2014).

Brounen, D. and N. Kok (2011), "On the economics of energy labels in the housing market", *Journal of Environmental Economics and Management*, Vol. 62, No. 2, pp. 166-179.

Copenhagen Economics (2012), *Multiple Benefits of Investing in Energy-efficient Renovation of Buildings – Impact on Finances*, report commissioned by Renovate Europe, Renovate Europe, Brussels, www.renovate-europe.eu/uploads/Multiple%20benefits%20of%20EE%20renovations%20in%20 buildings%20-%20Full%20report%20and%20appendix.pdf (accessed 23 June 2014).

ECF (European Climate Foundation) (2011), *Power Perspectives 2030: On the Road to a Decarbonised Power Sector*, ECF, Brussels, www.roadmap2050.eu/attachments/files/PowerPerspectives2030_FullReport.pdf (accessed 28 June 2014).

Ecofys (2013), *Saving Energy: Bringing Down Europe's Energy Prices for 2020 and Beyond*, report commissioned by Friends of the Earth Europe and Climate Action Network Europe, Ecofys, Utrecht, www.ecofys.com/files/files/foe-ecofys-2013-saving-energy-2020-and-beyond.pdf (accessed 22 June 2014).

EECI (Energy Efficient Cities Initiative) (2011), *Good Practices in City Energy Efficiency: Vienna, Austria (European Union) – Municipal Eco-Purchasing*, Washington DC, published online at www.esmap.org/sites/esmap.org/files/Vienna_Eco_Buy_final_edited_11-9-11.pdf (accessed 21 June 2014).

EECI and ESMAP (Energy Sector Management Assistance Program) (2014), *Good Practices in City Energy Efficiency: Vienna, Austria (European Union) – Municipal Eco-Purchasing*, published online at www.esmap.org/node/1315.

Eichholtz, P., Kok, N. and J. M. Quigley (2010), "Doing Well by Doing Good? Green Office Buildings", *American Economic Review*, Vol. 100, No. 5, pp. 2492-2509.

ENE (Environment Northeast) (2012), *Energy Efficiency: Engine of Economic Growth in Eastern Canada*, ENE, Ottawa, www.env-ne.org/public/resources/ENE_EnergyEfficiencyEngineofEconomicGrowth_EasternCanada_EN_2012_0611_FINAL.pdf (accessed 23 June 2014).

ExternE (2011), EcoSenseLE, EcoSenseLE website, EC DG Research, EC, Brussels, http://ecoweb.ier.uni-stuttgart.de/ecosense_web/ecosensele_web/frame.php (accessed 28 June 2014).

German Federal Ministry for the Environment (2012), *GreenTech made in Germany 3.0, Environmental Technology Atlas for Germany*, Public Relations Division, BMU (Nature Conservation and Nuclear Safety), Federal Ministry for the Environment, Berlin, www.bmub.bund.de/fileadmin/Daten_BMU/Pools/Broschueren/greentech_3_0_en_bf.pdf (accessed 21 June 2014).

Hoffman, R. (2009), "China's VAT system", *Beijing Review*, No. 31, China International Publishing Group, Beijing, www.bjreview.com.cn/business/txt/2009-08/03/content_210354.htm (accessed 28 June 2014).

Hyland, M., R. Lyons, and S. Lyons (2013), "The value of domestic building energy efficiency – evidence from Ireland", *Energy Economics*, Vol. 40(C), Elsevier, pp. 943-952.

IEA (International Energy Agency) (2014), *Energy Subsidies*, World Energy Outlook online resources, OECD (Organisation for Economic Co-operation and Development)/IEA, Paris, www.iea.org/publications/worldenergyoutlook/resources/energysubsidies (accessed 2 February 2014).

IEA (2013), *Energy Efficiency Market Report 2013: Market Trends and Medium-Term Prospects*, OECD/IEA, Paris, www.iea.org/w/bookshop/add.aspx?id=460 (accessed 19 June 2014).

IEA (2012a), *World Energy Outlook 2012*, OECD/IEA Paris, www.worldenergyoutlook.org/publications/weo-2012 (accessed 18 June 2014).

© OECD/IEA, 2014.

IEA (2012b), *Mobilising Investment in Energy Efficiency: Economic Instruments for Low-Energy Buildings*, Insights Paper, OECD/IEA, Paris.

James, K. (2011), "Exploring the Origins and Global Rise of VAT", *The VAT Reader*, Tax Analysts, Falls Church, www.taxanalysts.com/www/freefiles.nsf/Files/JAMES-2.pdf/$file/JAMES-2.pdf (accessed 22 June 2014).

Kronenberg, T., W. Kuckshinrichs and P. Hansen (2012), *Macroeconomic Effects of the German Government's Building Rehabilitation Program*, MPRA (Munich Personal RePEc Archive) Paper 38815, University Library of Munich, Munich, http://mpra.ub.uni-muenchen.de/38815/1/MPRA_paper_38815.pdf (accessed 23 June 2014).

Kuckshinrichs, W., T. Kronenberg and P. Hansen (2013), "Impact on public budgets of the KfW promotional programmes "Energy-efficient construction", "Energy-efficient refurbishment" and "Energy-efficient infrastructure" in 2011", STE Research Report, Forschungszentrum Juelich, commissioned by KfW Bankengruppe, Frankfurt, https://www.kfw.de/Download-Center/Konzernthemen/Research/Research-englisch/PDF-Dateien-STE-Reports/STE-Research-Report-April-2013-EN.pdf (accessed 22 June 2014).

Lehr, U., C. Lutz and M. Pehnt (2012), *Volkswirtschaftliche Effekte der Energiewende: Erneuerbare Energien und Energieeffizienz* (*Economic impacts of the energy transition [Energiewende]:*

renewable energies and energy efficiency), GWS and IFEU, Heidelberg, https://www.ifeu.de/energie/pdf/volkswirtschaftl_%20effekte_%20energiewende_broschuere_pehnt_RZ.pdf (accessed 28 June 2014).

Liddell, C. and C. Morris (2010), "Fuel poverty and human health: A review of recent evidence", *Energy Policy*, Vol. 38, No. 6, Elsevier Ltd., Amsterdam, pp. 2987-2997, www.sciencedirect.com/science/article/pii/S0301421510000625 (accessed 28 June 2014).

Meyer, A. and T. Johnson (2008), *Energy Efficiency in the Public Sector – A Summary of International Experience with Public Buildings and Its Relevance for Brazil*, ESMAP Report, World Bank, Washington DC, www-wds.worldbank.org/external/default/WDSContentServer/WDSP/IB/2012/06/21/000426104_20120621095049/Rendered/PDF/699140v20ESW0P0S0Part020May02902008.pdf (accessed 22 June 2014).

Ministère de l'économie et des finances (2013), *Les Recettes* fiscales, *Le budget* et *Les comptes de l'état*, Ministère de l'économie et des Finances website, http://www.performance-publique.budget.gouv.fr/budget-comptes-etat/budget-etat/approfondir/recettes-etat/recettes-fiscales#.U9jAt6Bqk-A, http://www.performance-publique.budget.gouv.fr/budget-comptes-etat/comptes-etat#.U9jBmaBqk-A and www.performance-publique.budget.gouv.fr/budget-comptes-etat/budget-etat#.U7m99vldUYM.

New Zealand Treasury (2008), *Public Sector Discount Rates for Cost-Benefit Analyses*, New Zealand Treasury, Wellington, www.treasury.govt.nz/publications/guidance/planning/costbenefitanalysis/discountrates (accessed 28 June 2014).

OECD (Organisation for Economic Co-operation and Development) (2014), OECD/European Environment Agency (EEA) database on instruments used for environmental policy and natural resources management, OECD Publishing, Paris, www2.oecd.org/ecoinst/queries (accessed 10 February 2014).

OECD (2013), *An OECD-Wide Inventory of Support to Fossil Fuel Use and Production*, OECD/Publishing, Paris, www.oecd.org/site/tadffss/Fossil%20Fuels%20Inventory_Policy_Brief.pdf (accessed 28 June 2014).

OMB (Office of Management and Budget) (2012), *Fiscal Year 2013 – Cuts, Consolidations, and Savings – Budget of the United States Government*, Executive Office of the President of the United States, Washington DC, www.whitehouse.gov/sites/default/files/omb/budget/fy2013/assets/ccs.pdf (accessed 28 June 2014).

PePS (Promoting an Energy-Efficient Public Sector) (2014), *Why Public Sector?*, webpage, PePS website, www.pepsonline.org/why.html (accessed 25 February 2014).

Prognos (2013), *Ermittlung der Wachstumswirkungen der KfW-Programme zum Energieeffizienten Bauen und Sanieren* [*Determination of the growth effects of the KfW programmes for energy-efficient building and reorganisation*], Prognos, Berlin, Basel, www.kfw.de/Download-Center/Konzernthemen/Research/PDF-Dokumente-alle-Evaluationen/Wachstumseffekte-EBS-Endbericht.pdf (accessed 22 June 2014; in German).

© OECD/IEA, 2014.

Pollitt, H. (2014), Personal communication.

PwC (PricewaterhouseCoopers), Significant and Ecofys (2009), *Collection of Statistical Information on Green Public Procurement in the EU – Report on Data Collection Results*, PwC, London, http://ec.europa.eu/environment/gpp/pdf/statistical_information.pdf (accessed 22 June 2014).

Roland-Holst, D. (2011), *Driving California's Economy: How fuel Economy and Emissions Standards will Impact Economic Growth and Job Creation*, Research Paper 1103011, University of Berkeley, California, http://next10.org/sites/next10.huang.radicaldesigns.org/files/Final_vehicle_efficiency_report.pdf (accessed 25 February 2014).

Sauter, R. and A. Volkery (2013), *Review of Costs and Benefits of Energy Savings*, report by the IEEP (Institute for European Environmental Policy) for the Coalition of Energy Savings, Task 1 Report, Brussels, www.ieep.eu/assets/1267/Energy_Savings_2030_IEEP_Review_of_Cost_and_Benefits_of_Energy_Savings_2013_published.pdf (accessed 22 June 2014).

SEAI (Sustainable Energy Authority of Ireland) (2013), *Energy Savings in the Public Sector*, SEAI, Dublin, www.seai.ie/Publications/Your_Business_Publications/Public_Sector/Energy_Use_in_the_Public_Sector.pdf (accessed 28 June 2014).

UK HM Treasury (United Kingdom, Her Majesty's Treasury) (2003), *The Green Book: Appraisal and Evaluation in Central Government*, United Kingdom government, London.

US EPA (United States Environmental Protection Agency) (2010), *Evaluation of Conservation measures for Waste Water Facilities*, Case Study 7, Report EPA 832-R-10-005, US EPA, Washington DC, http://water.epa.gov/scitech/wastetech/upload/Evaluation-of-Energy-Conservation-Measures-for-Wastewater-Treatment-Facilities.pdf (accessed 28 June 2014).

Van Wie McGrory, L. et al. (2002), "Market Leadership by Example: Government Sector Energy Efficiency in Developing Countries", in *Proceedings, 2002 Summer Study on Energy Efficiency in Buildings*, American Council for an Energy-Efficient Economy, Washington DC.

WHO (World Health Organization) (2011), *Health in the Green Economy: Health Co-benefits of Climate Mitigation: Transport Sector*, WHO, Geneva, www.who.int/hia/examples/trspt_comms/hge_transport_lowresdurban_30_11_2011.pdf?ua=1 (accessed 28 June 2014).

World Bank (2014), Energy Efficient Cities Case Studies Database, Washington DC, www.esmap.org/node/231 (accessed 21 June 2014).

World Bank (2011), *Public Procurement of Energy Efficiency Services: Lessons from International Experience*, The World Bank, Washington DC, available at www.esmap.org/sites/esmap.org/files/P112187_GBL_Public%20Procurement%20of%20Energy%20Efficiency%20Services_Lessons%20from%20International%20Experience_Singh.pdf.

© OECD/IEA, 2014.

Chapter 4

Health and well-being impacts of energy efficiency

Key points

- *Aside from potential energy demand reductions, improving energy efficiency in buildings creates conditions that support improved health and well-being for occupants. Positive health outcomes are consistently strongest among vulnerable groups, including children, the elderly and those with pre-existing illnesses.*
- *The most prominent health impacts associated with energy efficiency improvements include reduced respiratory disease symptoms and lower rates of excess winter mortality (EWM) in cold climates. Fewer deaths from dehydration are reported in heat extremes.*
- *Recent evidence shows that chronic thermal discomfort and fuel poverty have negative mental health impacts (anxiety, stress, depression and worry about physical health); energy efficiency improvements can improve mental well-being.*
- *Health improvements at the individual level generate indirect social impacts and relieve pressure on public health budgets. Modelling of a high energy efficiency scenario showed that reduced indoor air pollution could save the European public health budget USD 99 billion per year in 2020.*
- *Overlaying proven metrics and assessment methods from epidemiological disciplines with financial metrics can generate market values for identified health benefits, enabling these outcomes to be built into robust policy assessment frameworks.*
- *When quantified health and well-being impacts are included in assessments of energy efficiency retrofit programmes, the benefit-cost ratio can be as high as 4:1, with health benefits representing up to 75% of overall benefits.*
- *The body of evidence linking improved health and well-being to energy efficiency measures has prompted several governments to make addressing fuel poverty a central element of energy policy, often optimising investments by targeting vulnerable groups.*

Introduction: Linking energy efficiency and health

The link between healthy environments and healthy people is well known. More recently, clear evidence has emerged that energy efficiency measures can support good health and offset the potential negative outcomes associated with poor quality buildings and indoor environments. Health and well-being are products of complex interactions among different physical, social, economic and environmental factors that can trigger and offset the potential health outcomes of energy efficiency measures (Maidment et al., 2014; Murray et al., 2000).

While energy efficiency measures in diverse sectors show potential to deliver health and well-being improvements, this chapter focuses on the most straightforward and compelling case: energy efficiency measures in buildings, where the impacts on occupants are easier to verify. The chapter points to important impacts in other sectors and those associated

© OECD/IEA, 2014.

with reducing outdoor air pollution (Box 4.4). In 2012, 7 million deaths (one in eight global deaths) resulted from indoor and outdoor air pollution, making this the world's largest single environmental risk (WHO, 2014).

A whole-building approach to energy efficiency – including improving insulation, heating and cooling systems, lighting and energy-using equipment – can reduce energy costs and support warmer, drier, more comfortable indoor environments. Both outcomes have been linked with direct impacts such as lowered risk of illness and with indirect physical health benefits (Thomson et al., 2013). But evidence goes beyond physical health: energy efficiency measures can alleviate the mental health burdens often associated with cold homes, including the financial stress of high energy bills (Liddell and Guiney, 2014). Conversely, poorly executed energy efficiency interventions that degrade indoor air quality carry a risk of damaging health. This risk must be carefully monitored to ensure that health impacts remain positive.

When quantified or valued from qualitative perspectives, the health and well-being benefits of well-executed energy efficiency programmes are shown to outweigh the benefits of energy demand reduction – in terms of both social and financial value. Moreover, cost-effectiveness analysis (CEA) indicates that such programmes generate cost savings for the public health sector that outweigh the initial expenditure on the energy efficiency intervention.

Methods for measuring and quantifying health and well-being impacts in ways that support policy development are developing rapidly. These techniques are helping to build understanding about the causal pathway between improving indoor environments and health and well-being outcomes. The values, methods and issues presented in this chapter provide strong evidence that underpins a growing consensus on the important role of energy efficiency in delivering health and well-being benefits. They also provide a point of departure for further investigation needed to make the multiple benefits[1] approach mainstream when energy efficiency and health goals overlap.

Targeting energy efficiency policies to address fuel poverty

In industrialised countries, the term "fuel poverty" was coined to identify individuals or families who technically have access to energy but cannot afford adequate levels to meet their basic needs. Fuel poverty is broadly defined as representing any household spending more than 10% of its annual income on energy (Boardman, 2010) or, more recently in England, as a relative measure of energy costs compared with the median equivalised energy bill in the population (Hills, 2012). Most often, fuel poverty arises at the nexus of low income, poor housing quality and high energy costs. Recent estimates suggest more than 150 million people are living in fuel poverty in the European Union alone (Bird, Campbell and Lawton, 2010) – and that this number is growing.

Health and well-being benefits of energy efficiency improvements are consistently strongest among vulnerable groups; children in low-income families, the elderly and those with pre-existing illnesses are most at-risk to the health impacts of fuel poverty. Governments use a range of policies to mitigate fuel poverty, including support payments for fuel costs, social tariffs (subsidies) on energy prices, grant schemes for expenses associated with building or equipment energy efficiency upgrades, or free retrofit programmes. To date, programmes for energy efficiency retrofitting of low-income housing have delivered the greatest benefits, with health improvements representing as much as 75% of the total return on the investment for these interventions (Grimes et al., 2011). Energy efficiency retrofits offer a

1 In other literature, these impacts have been variously labelled "co-benefits", "ancillary benefits" and "non-energy benefits" (NEBs) – and are often used interchangeably with "multiple benefits". The International Energy Agency (IEA) uses the term multiple benefits, which is broad enough to reflect the heterogeneous nature of outcomes and to avoid pre-emptive prioritisation of various benefits; different benefits will be of interest to different stakeholders.

© OECD/IEA, 2014

more enduring solution than subsidies and grants by addressing the cause of fuel poverty, rather than the symptoms.

Governments in several IEA member countries have launched targeted efforts to address the difficulties of fuel-poor households, often by explicitly integrating the potential for health benefits into energy efficiency policies. Frontrunner countries that are taking a strategic approach to tackling fuel poverty include Ireland, the United Kingdom and the United States (Box 4.1).

Box 4.1 IEA governments get serious about energy efficiency policies that target health and well-being

Almost all IEA member countries have programmes underway to support energy efficiency building retrofits; to date, only a few have sought to use policy design to maximise the potential health and well-being benefits of these measures.

Assessment of the health impacts achieved through Irish energy efficiency policy has convinced the Department of Energy and Natural Resources of the significance of the health benefits approach (Clinch and Healy, 2001). Ireland's *Affordable Energy Strategy 2011* recognises the critical link between energy and health, including over 40 separate references to health and well-being issues. The *Better Energy: Warmer Homes** scheme aims to deliver these benefits to low-income households as part of efforts to tackle high fuel poverty and rising public spending on general household benefits and fuel subsidies (Scheer, 2013). Ireland's Department of Health and Children is gradually becoming alert to the role of energy efficiency in addressing health challenges.

On a grant-tiered basis, the *Warm Up New Zealand: Heat Smart*** (*WUNZ: HS*) programme provided energy efficiency retrofits to households on low and medium incomes. Although administered by the Energy Efficiency and Conservation Authority, WUNZ: HS was a strong example of multiple policy objectives being integrated within a common policy. As a result of more than a decade of research by *He Kainga Oranga* (the Housing and Health Research Programme of Otago University, New Zealand), health outcomes were factored into the programme design and New Zealand's 2011 Energy Strategy included a stated objective of "warm, dry and energy efficient homes with improved air quality to avoid ill-health and lost productivity" (NZ MED, 2011). The significant positive health impacts of this programme are described in Box 4.2.

In England, the *Warm Front**** scheme provided insulation and heating systems to over 2 million vulnerable households since 2000. This activity has been driven by statutory targets to address fuel poverty. While neither the *Warm Front* scheme nor the Annual Energy Statement mentioned health, United Kingdom (UK) evaluations provide a strong body of evidence showing how local energy efficiency programmes influence health among the fuel poor (Marmot Review Team, 2011).

In the United States, more than 7 million low-income homes have been weatherised over three decades via the United States (US) Department of Energy's *Weatherization Assistance Program* (*WAP*)****, which provides grants to local authorities to improve the energy efficiency of the homes of low-income families. The *Low Income Home Energy Assistance Program* (*LIHEAP*) carries out similar initiatives specifically targeted to low-income families. The Oak Ridge National Laboratory (ORNL) is tasked with ongoing evaluation of the multiple benefits of weatherisation, including the impacts on health and well-being (Schweitzer and Tonn, 2003).

At the European level, the Third Energy Package of the European Union includes specific provisions for providing assistance to vulnerable energy customers (EC VCWG, 2013). The forthcoming recast of the European Energy Performance of Buildings Directive (EPBD; Directive 2010/31/EU) may present an opportunity to build the issue of health and well-being impacts of energy efficiency into the requirements it sets for member states.

* www.seai.ie/Grants/Warmer_Homes_Scheme/.
** www.eeca.govt.nz/eeca-programmes-and-funding/programmes/homes/insulation-programme.
*** www.gov.uk/warm-front-scheme. *Warm Front* ended January 2013 and has been replaced by the generally available *Green Deal*, coupled with *Energy Company Obligations* to provide energy-saving measures for hard-to-treat homes and customers most in need.
**** www.energy.gov/eere/wipo/weatherization-assistance-program.

© OECD/IEA, 2014.

Surveys of 2 500 participants in the United Kingdom's *Warm Front* programme found that providing insulation and efficient central heating systems to low-income households led to increased indoor temperatures and improved thermal comfort (Hong et al., 2009). Analysis of around 4 000 homes participating in the same programme concluded that increased temperatures are linked to improved health and fewer excess winter deaths (Gilbertson, Grimsley and Green, 2012).

Fuel poverty is also strongly associated with sub-optimal mental health driven, in particular, by the financial stress of coping with high energy bills and debt. Energy efficiency measures that improve the affordability of energy bills in low-income homes can have a measurable effect on improving mental well-being (e.g. happiness and coping) and preventing mental disorders (e.g. anxiety and borderline depression) (Liddell, Morris and Langdon, 2011).

Targeting energy efficiency measures to households most in need, particularly those in fuel poverty, can both maximise benefit-cost ratios and contribute to improving social equality. Various considerations, in particular the trade-off between administrative cost and improved outcomes, must to be taken into account when designing energy efficiency policies that target the health and well-being benefits for fuel-poor households (see discussion on "indirect impacts" below).

The range of health and well-being impacts

Energy efficiency has been linked to an extensive range of health benefits and to flow-on outcomes for psychosocial functioning in society. To understand the dynamics behind health outcomes, it is useful to separate the issue into two parts:

- The **exposure factors** linked to environmental conditions that are critical to improving health and well-being; and
- The potential **health and well-being outcomes** that could be expected as a result.

Evidence about the impact of energy efficiency on exposure factors is relatively strong; it is easily measured and has been a driver of energy efficiency polices for some time. Drawing direct causal links with health and well-being outcomes presents greater challenges; while clear correlations are evident, further investigation is needed to reinforce the emerging evidence. Both aspects are discussed below, but the novel aspect of the potential health and well-being outcomes that improved energy efficiency can deliver are the main focus of this chapter.

Exposure factors and the role of energy efficiency in mitigating their effects

Three key exposure factors related to indoor environments are shown to have major impacts on human health and well-being: i) thermal quality; ii) air quality; and iii) mould caused by dampness.

Thermal quality refers to whether the indoor temperature is comfortable and healthy.[2] While most evidence relates to the impact of cold environments, over-heating can also damage health through dehydration (Naughton et al., 2002). Energy efficiency retrofit programmes that include installing insulation are shown to enable occupants to raise indoor air temperatures to healthy levels. Temperature is the biggest driver of health and well-being impacts; among energy efficiency measures, insulation has the largest potential to drive health improvements (Box 4.2).

2 The World Health Organization (WHO) recommends healthy indoor air temperatures as between 18 degrees Celsius (°C) and 21°C.

© OECD/IEA, 2014.

Box 4.2 Improvements in building insulation deliver greatest health impacts

WUNZ: HS is a five-year government programme (budget of USD 300 million or NZD 347 million) that provided funding for insulation retrofits and clean, efficient heating grants for households. A detailed assessment of energy efficiency impacts on health, commissioned by the Ministry of Economic Development, highlighted broad social benefits and made a significant contribution to international understanding of this issue. This led to increased strategic co-operation between New Zealand's Energy Efficiency and Conservation Authority and the Ministry of Health.

The assessment, which involved two community trials, highlighted two significant causal links between energy efficiency measures and occupant health. In the first study, retrofitting of insulation in 1 400 households where occupants had respiratory problems led to clear improvements in occupants' respiratory and mental health (Howden-Chapman et al., 2007). The second trial assessed the impacts of combining retrofitted insulation and installation of non-polluting heating in 400 households where there was a child with doctor-diagnosed asthma, and also found a causal link between more efficient, non-polluting heaters and improvements in children's respiratory health and school attendance (Howden-Chapman et al., 2008).

In addition to generating robust empirical evidence about the causal links between energy efficiency measures and health, the assessment took the further step of monetising those impacts. The net present value (NPV)* of observed impacts was calculated on the basis of the avoided cost of hospital admissions and included into a benefit-cost analysis (BCA**). The monetised impacts (which included health and other NEBs as well as energy cost savings) delivered an overall benefit-cost ratio for the programme of over 4:1. Over a 20 year period, the insulation measures were expected to contribute USD 1 billion to USD 1.5 billion (NZD 1.1 billion to NZD 1.7 billion) in benefits. Results show that 99% of that figure is made up of health benefits, with reduced mortality accounting for 74% (Grimes et al., 2011).

Of these monetised benefits, the highest were achieved among recipients who held a Community Services Card (issued to families on low to modest incomes). The total annual benefit – energy and health – after retrofitting was USD 519 (NZD 613) for cardholders compared to USD 183 (NZD 216) for non-cardholders (Telfar et al., 2011). The programme also found that targeting dwellings in colder areas or with lower levels of insulation would deliver enhanced benefit-cost ratios.

* The use of an NPV calculation to identify the cost of past impacts in current values is explained further in the Companion Guide at the end of this publication.
** Although the commonly used term is "cost-benefit analysis" (CBA), the IEA prefers the term "benefit-cost analysis" (BCA) due to the fact that the ratios produced are expressed as "benefit:cost". The actual approach is the same.

Ensuring good indoor **air quality** means minimising levels of toxins such as radon (which is naturally occurring in some countries)[3] and other gases and particulate matter that can be generated by inefficient heating and cooking methods. Integrating ventilation measures into energy efficiency retrofit packages can mitigate exposure to these toxins.

Dampness can generate and aggravate a range of illnesses and plays a particularly strong role in inducing symptoms of allergies and respiratory disease, but also other health and well-being challenges. It can also lead to mould growth, which has independent effects on health. A US study estimated the cost of asthma induced by dampness and mould in homes at USD 3.5 billion per year (Mudarri and Fisk, 2007). Procedural safeguards must be built into energy efficiency interventions to ensure that measures, such as draught-proofing, do not hinder air flow; and to mitigate the risk of negative impacts on these factors created by inadvertently locking in toxins and damp.

3 Radon levels are problematic in the United States and in some European countries, such as the United Kingdom and Switzerland.

© OECD/IEA, 2014.

While individual measures can address each of the exposure factors, a holistic approach to energy efficiency building retrofitting delivers higher health and well-being impacts. Improving heating systems and draught-proofing contribute to thermal quality. Adding better ventilation systems can improve indoor air quality by reducing the incidence of toxic gases and particulates caused by inefficient heating systems (such as unflued gas and wood burners). In addition to contributing to improved air quality, measures to replace old inefficient cooking and refrigeration appliances can reduce the number of injuries caused by their use and support improved nutrition.

All of these measures, particularly ventilation, play a role in reducing indoor dampness and the associated build-up of mould that exacerbates many health conditions (Thomson et al., 2013). Additional recorded impacts from energy efficiency measures include increasing the useable space within the home as more rooms can be affordably heated and reducing energy bills, which has a positive impact on the mental well-being of occupants.

A growing body of evidence reveals clear pathways by which energy efficiency measures address exposure factors and indicates their potential role in generating health and well-being outcomes (Table 4.1). The list below, which is non-exhaustive, results from a thorough review of primary and secondary evidence sources and from discussion with experts.[4]

Complex dynamics are at work in the exposure-outcome pathway between energy efficiency and health and well-being outcomes, which makes it challenging to draw strong conclusions about the causal link between a particular change in one exposure factor and the measured outcomes for health and well-being. Several recent studies show evidence of the role energy efficiency improvements play in supporting health and well-being, particularly among vulnerable populations such as children, older people and those with existing illnesses.

The quality of the evidence varies with the sample sizes and robustness of methods used, but compelling quantitative and qualitative results have been replicated in a number of countries, providing insights into both the tangible and intangible benefits of energy efficiency. The following summary of results categorises the potential benefits as either direct or indirect according to the proximity of the causal link with originating energy efficiency measures.

Direct impacts

Direct health and well-being impacts have been the focus of most studies to date. These include physical health impacts from both hot and cold extremes such as reductions in mortality, symptoms of respiratory and cardiovascular diseases, allergies, arthritis and rheumatism, and injuries, as well as reduced circulation of infectious diseases through close contact. Of these, the strongest evidence of a causal link with energy efficiency is found in relation to temperature-related deaths, and respiratory and cardiovascular diseases. Evidence is strengthening that direct benefits also extend to mental well-being issues such as the reduction of stress and depression.

Physical health impacts from residential measures

A meta-analysis of studies investigating the health benefits of energy efficiency interventions (including insulation, heating system improvements and improved ventilation) cites a diverse range of positive physical health impacts including reduced susceptibility to illness, reduced incidence of heart attacks and strokes due to high blood pressure,

4 In particular, discussion held at the IEA-European Environment Agency Expert Roundtable on the Health and Well-being Impacts of Energy Efficiency, Copenhagen, Denmark, April 2013: www.iea.org/workshop/roundtableonthehealthwell-beingimpactsofenergyefficiencyimprovements.html.

© OECD/IEA, 2014.

Table 4.1 Overview of direct and indirect impacts of improved energy efficiency on health and well-being

Energy efficiency measures	Impacts associated with energy efficiency measures		Potential health outcomes - direct		Potential health outcomes - indirect	
Insulation	Warmer, drier, indoor environment	Comfortable temperature	Reduced deaths from cold and hot spells^{+++}	Reduced excess (winter and summer) mortality^{+++}	Reduced absenteeism from school^{++}	
Draught-proofing, pipe lagging, lighting			Reduced symptoms of respiratory disease: asthma, lung cancer, Chronic Obstructive Pulmonary Disease^{+++}		Improved academic performance^{+}	
Extractor fans	Well ventilated/ good air quality	Reduced damp*	Reduced symptoms of cardiovascular disease (e.g. angina, atrial fibralation, risk of stroke)$^{+++}$	Reduced hospitalisation^{++}	Reduced absenteeism from work^{++}	
		Reduced mould*	Reduced depression^{++}		Increased productivity^{+}	
			Reduced arthritis and rheumatism^{++}		Increased earning power^{++}	
			Reduced injuries and death^{+}			
Efficient, effective heating systems		Comfortable temperature	Reduced allergies^{++}	Reduced pharmaceuticals $^{+}$		Reduced public and private spending on health
		Reduction of gas and particulates* $^{+++}$	Reduced respiratory disease: asthma, lung cancer, Chronic Obstructive Pulmonary Disease^{+++}	Reduced hospitalisation^{+++}		
			Reduced injuries and death^{++}			
		Increased usable living space	Reduced stress^{++}		Increased sociability^{+}	
			Reduced close contact infectious diseases^{++}		Increased space for homework^{+}	
Efficient and effective cooking/ refrigeration systems		Reduced gas and particulates*	Reduced injuries and death^{+}			
		Improved fitness for purpose (i.e. better refrigeration and cooking facilities)	Improved nutritional status^{++}			
	Reduced energy bills/ reduced exposure to energy price fluctuations	Increased sense of control^{+}	Reduced stress and despression^{++}			
		Less fear of falling into debt^{+}				
		More disposable income	Increased purchase of food and other essentials^{+}		Improved nutrition^{++}	
					Increased access to preventative health care^{+}	

Notes: This graphic illustrates the impact pathways from energy efficiency measures to three major impacts. Colour coding established in the impacts column corresponds with the various outcomes a measure could generate for health. This simplified flow diagram does not depict all of the complex interrelationships related to energy efficiency and health and well-being outcomes.

$^{+}$, $^{++}$, or $^{+++}$ symbol indicates the strength of the evidential basis, with $^{+}$ being lowest and $^{+++}$ being highest.

* Caution: Sealing homes without adequate ventilation can cause unintended negative consequences for health.

Source: Unless otherwise noted, all material in figures and tables in this chapter derives from IEA data and analysis.

© OECD/IEA, 2014.

improvements to respiratory health (particularly asthma symptoms in children), reduced injuries and reduced excess seasonal mortality (Maidment et al., 2014).

Energy efficiency measures are shown to have the greatest impact on improving general health (Platt et al., 2007; Kearns and Petticrew, 2008) and respiratory health (Howden-Chapman et al., 2007). Where energy efficiency generates a temperature gain of even 1°C or 2°C in the home, this can lead to improvements in childhood asthma, reducing wheezing, coughing and lower respiratory tract symptoms, as well as fewer visits to doctors and pharmacists (Howden-Chapman et al., 2008). BCA of energy efficiency policies carried out by the Sustainable Energy Authority of Ireland indicates that greater comfort and reduced excess winter mortality and morbidity represent as much as 70% to 80% of the overall benefits (Scheer, 2012).

Cold temperatures

In cold climates, energy-inefficient housing is an important driver of EWM – the increased rate of death witnessed in most countries during winter periods. Within most member countries of the Organisation for Economic Co-operation and Development, EWM ranges between 5% and 30% (Clinch and Healey, 2001). In temperate countries such as Ireland, Scotland and New Zealand, and even in southern European countries such as Greece and Portugal, EWM rates tend towards the higher end of this scale. This has been linked to inadequate housing and heating, and a failure to protect many older people from extreme winter weather (Healy, 2003). Studies conducted in New Zealand and Ireland – two countries with temperate climates – estimate that poorly insulated housing is responsible for an average of 1 700 excess winter deaths from cold in each country annually (Howden-Chapman et al., 2012; Clinch and Healy, 2001).[5]

In cold conditions, improving the affordability of heating has been linked to significantly higher weight-for-age scores among infants, lower malnutrition risk and advanced developmental status (Frank et al., 2006).

Hot temperatures

Energy inefficient housing has also been implicated in excess summer deaths (Grynszpan, 2003), suggesting that energy efficiency measures could also help to guard against the impact of heat extremes. Because of the nature of the in-built temperature management systems in the human body, heat and subsequent dehydration is in fact a greater contributor to temperature-related deaths than cold. The heat wave that struck France in 2003 resulted in an excess of 14 800 deaths between 1 and 20 August, mostly of elderly individuals (Vandentorren et al., 2012). Excessive heat has particular negative health impacts for pregnant women (Liddell, 2014); studies from diverse countries have linked excess heat to lower birth weight of babies.

The same energy efficiency measures that keep heat inside in cold temperatures, will keep heat out when outdoor temperatures are high. Insulation and energy efficient windows will prevent outdoor heat from penetrating the building, while deeper energy efficiency principles, such as building positioning and depth of eaves, can be built in at design stages to facilitate natural cooling.

The IPCC Working Group Three recently restated the increased risk of morbidity and mortality during periods of extreme heat – particularly for vulnerable urban populations. While it has been suggested that climate change could have a positive effect on winter mortality as average temperatures rise, the matter is under debate (Christidis, Donaldson

5 In New Zealand, it was estimated that poorly insulated housing was responsible for an average of 1 600 excess winter deaths annually (Howden-Chapman, 2012). The equivalent figure for Ireland was estimated at between 1 500 and 2 000 in 2000 (Clinch and Healy, 2000).

© OECD/IEA, 2014.

and Stott, 2010; Staddon, Montgomery and Depledge, 2014). Retrofitting existing housing and raising the standards of new housing can help occupants adapt to weather extremes as well as mitigate the effects of climate change by reducing energy consumption.

Mental health

Empirical evidence of the impact energy efficiency improvements can have on adult mental health[6] has emerged in the last ten years. Relevant studies show significant and consistently reduced stress and depression, and improved levels of well-being (Liddell and Morris, 2010). A flow-on impact of prevention of physical illness is also evident (Marmot Review Team, 2011).

Inefficient housing and fuel poverty have been shown to affect mental health by way of chronic thermal discomfort (Gilbertson, Grimsley and Green, 2012), impacts of condensation, damp and mould (Liddell, 2013a), financial stress related to high energy bills (Anderson et al., 2000) and the experience or fear of falling into debt (Tod et al., 2012). Alleviation of financial stress is the most commonly observed driver of mental health improvements through energy efficiency. Studies point to an exponential effect of multiple stress factors as driving cumulative stress effects (Liddell and Guiney, 2014), with other common stressors including: people's concern that cold is damaging their physical health (De Haro and Koslowski, 2013); the effect of "spatial shrink" from living in only one or two rooms that can be affordably heated (Liddell and Morris, 2010; Howden-Chapman et al., 2009); social stigmatisation within one's community (Clare, 2013); damage to possessions from damp and mould (Liddell, 2013b); and the absence of any sense of control over the problem (Stearn, 2012).

When energy efficiency measures reduce energy-related expenses, they can drive a further set of health benefits, both directly (e.g. by reducing stress) and indirectly (e.g. by freeing up finances for other health-supporting expenditures). Energy bill-related benefits are observed in households in fuel poverty in particular, where unmanageable energy bills have far-reaching consequences. The potential of energy efficiency measures to reduce energy bills depends on the financial structure of the energy efficiency intervention; i.e. whether and how the up-front investment falls to the occupant for repayment. In what is known as the rebound effect, recipients are likely to make a trade-off between improving their indoor environment and reducing their energy expenses, which will affect the benefits that could result (Rebound effect perspective 4).

Following energy efficient upgrades, studies show statistically significant reduction in stress among low-income households, associated with lowered energy bills and improved ability to pay (Box 4.3). For example, the installation of thermostatic controls on heating systems can ease stress by enabling householders to self-manage the trade-off between warmth and bills (Bashir et al., 2014).

Efforts to test the link between energy efficiency improvements and mental well-being have applied standardised surveys to generate robust evidence, leading to results of a high standard. Consequently, poor housing (generally) and fuel poverty (in particular) are now both listed as key factors associated with mental health risk in Europe. Experts argue that sufficient evidence exists to draw a direct link between the diverse stressors associated with fuel poverty and sub-optimal mental well-being (Liddell and Guiney, 2014), and to support the inclusion of mental health impacts into the assessment of energy efficiency policy impacts.

6 Mental well-being has been defined as "a dynamic state that refers to individual's ability to develop their potential, work productively and creatively, and build strong and positive relationships with others and contribute to their community" (RCP, 2011).

© OECD/IEA, 2014.

Rebound effect perspective 4

Analysis of health and well-being impacts

The potential for health and well-being impacts from energy efficiency improvements arises independently of any energy saving they will generate. Whether simple insulation measures or a deep weatherisation package, the benefits of an improved indoor environment will be available to all occupants.

The opportunity for energy saving – and its impacts – should not be overlooked. Lowered energy bills can also lead to health and well-being benefits, particularly for households in fuel poverty.

While energy demand reduction is achieved in some cases, some building occupants may choose to take increased service instead of saving energy, for example by choosing to increase heating in a previously under-heated home because bill-paying occupants feel the money is now better spent (Howden-Chapman et al., 2012). The unrealised energy saving forms part of a rebound effect, and can be as high as 10% to 30% of projected energy demand reduction (Sunikka-Blandk and Galvin, 2012).

Arguably, the success of the energy efficiency intervention is in providing building occupants with this choice, and each occupant is free to resolve the situation by trading off the available benefits according to personal needs. A rebound effect driven by an occupant's choice to increase comfort and improve the conditions for health and well-being is unlikely to be considered as a negative result overall.

Box 4.3 UK study produces concrete results on the mental health impacts of energy efficiency

With multiple objectives, including tackling fuel poverty and creating jobs, and a total budget of USD 40 million* over three years, the *Kirklees Warm Zone* project provided free energy retrofits across the entire borough of Kirklees, England. It was ground-breaking for many reasons including its unprecedented scope and scale, its novel approach to delivery, and the concrete results it produced about the impact of energy efficiency on mental health. The intervention consisted of insulation and heating system upgrades as well as additional minor measures.

Anticipating that significant physical health impacts were unlikely to be observed within the short period following energy retrofits, project evaluators focused instead on mental health benefits among adults. They applied standardised surveys already used by the mental health profession to measure the impact of energy efficiency improvements on common mental disorders, including odds ratios** of self-reported mental health disorders, derived or adapted from published studies. The odds ratios were then applied to the baseline prevalence of common mental disorders to determine the reduced incidence of mental disorders. These results were converted to quality-adjusted life years (QALYs), a standardised measure in epidemiology (see discussion on "methodological approaches" below).

An overall benefit-cost ratio of 0.2:1 was calculated for the whole project, indicating that 20% of all programme costs were recouped on the basis of positive health impacts. Mental well-being impacts accounted for approximately 50% of that benefit (Liddell, Morris and Langdon, 2011).

* Converted from GBP 24 million at a rate of GBP 1 = USD 1.68435.
** An odds ratio presents the odds of an event occurring in one group to the odds of it occurring in another group; it can be used to quantify hard-to-measure impacts (see more detailed explanation in the Companion Guide at the end of this publication).

© OECD/IEA, 2014.

While the evidence base is of a high standard, this is a new area of investigation and robust studies are sparse. In particular, greater clarity is needed about the construct validity of the different measurement scales used in the context of energy efficiency impacts.

Indirect impacts

The initial health impacts arising from energy efficiency measures also stimulate indirect impacts for both the individual and society. Drawing the causal link of these benefits tends to be more difficult, but they remain important to understanding the full potential that energy efficiency might deliver to the individual and can aid in the process of monetising benefits. Impacts for society as a whole, primarily in terms of reduced public spending on health and social welfare services, are of major significance for broader policy decision making.

Public health spending

Health and well-being improvements have significant impacts for society as a whole, some of which are reflected in savings in public health spending. In an assessment of energy efficiency measures in Europe out to 2020, health benefits from reduced indoor air pollution are shown to be worth USD 45 billion to USD 99 billion (EUR 33 billion to EUR 73 billion) annually in a low energy efficiency scenario and as much as USD 87 billion to USD 190 (EUR 64 billion to EUR 140 billion) billion annually in a high energy efficiency scenario (Copenhagen Economics, 2012). These savings derive from values for improved life quality, lowered public health spending and fewer missed days of work. Continuing the proposed investments after 2020 may double the total value.

Estimates of public health spending are highly dependent on the health system in each country as well as on climatic factors. As evidence of impacts at the level of the individual increases, these estimates will be refined. Greater certainty of the health impacts of outdoor air pollution pitches the annual health benefit of reducing air pollution in Europe at USD 7 billion to USD 11billion (EUR 5 billion to EUR 8 billion) (Copenhagen Economics, 2012).

Other indirect impacts

Energy efficiency improvements can, through their impact on health, also indirectly provide other benefits. Growing evidence supports a significant relationship between improving the quality of indoor environments (in terms of air quality and lighting) and enhanced cognitive ability and productivity in non-domestic environments, such as schools, businesses and offices. Additional indirect impacts on productivity and family cohesion are evident in home environments as well.

The positive impact of energy efficiency measures (e.g. installation of controls, heating and ventilation) on the work environment contributes to substantial economic benefits (Fisk, Black and Brunner, 2011). Improving the learning ability of students in primary schools is estimated to potentially boost gross domestic product (GDP) in Denmark by USD 235 million (EUR 173 million) through the increased productivity of future workers (Slotsholm, 2012). Energy efficiency in workspaces can reduce the prevalence of communicable illness, generating fewer days lost due to illness[7] while also enhancing occupant satisfaction. In evaluating a wide range of buildings, the European Health Optimisation Protocol for Energy-efficient buildings project found that low-energy buildings have lower rates of reported ill health and absenteeism due to poor indoor air quality (HOPE Project, 2005).

Energy efficiency measures that result in lower energy bills can help to address the "heat or eat" dilemma faced by some low-income households. This refers to a trade-off made

7 The "No Lift Days" initiative implemented in government buildings by ENEA, the Italian National Agency for Technologies, Energy and Sustainable Economic Development, has generated statistically significant improvement in peak arterial pressure among employees (Delussu, 2014).

© OECD/IEA, 2014.

between paying for energy bills and other necessities, such as food. Nutritional problems (both malnutrition and obesity) that may arise from this trade-off can be addressed by energy efficiency measures (Cook et al., 2008).

Finally, some evidence suggests that poor housing can lead to or exacerbate health and social problems. Crowding into a single heated room increases exposure to communicable disease and also limits ability to do homework (and therefore educational achievement) or other private tasks, which can create tension within the household. Energy efficiency retrofits can address issues of social isolation arising from occupants of poor quality housing being embarrassed by their uncomfortable conditions (Barton, Basham and Shaw, 2004; Bashir et al., 2014). Other flow-on effects within the local community have been hypothesised but require further research; these include improved social cohesion and sense of community among residents, higher rates of school attendance, healthier lifestyles and improved access to local services (Dempsey et al., 2011)

Safeguarding against negative health impacts of poorly executed energy efficiency measures

Improving energy efficiency in buildings generally improves the indoor environment. But if energy efficiency measures are implemented incorrectly, they can have negative impacts on indoor air quality and thus on health and well-being.

Increasing air-tightness – through insulating, sealing and draught-proofing homes – without ensuring good ventilation systems[8] will reduce air exchange rates and generate a negative impact on indoor air quality (Shrubsole et al., 2014). This can raise humidity levels and lead to mould and dust mites, prompting allergic symptoms and asthma (ECA, 2013). Reduced air exchange can also lead to dangerous build-up of radon or carbon monoxide generated by unflued solid fuel or gas heaters, as well as other pollutants such as second-hand tobacco smoke (Davies and Oreszczyn, 2012), all of which have been strongly associated with lung cancer. Each year, 22 000 radon-related lung cancer deaths are reported in the United States – more than from drunk driving accidents – costing approximately USD 1.1 million per case (US EPA, 2011). The combined health impacts of poor indoor air quality in Europe have been estimated at approximately 2.2 million disability-adjusted life years (DALYs)[9] per year, which would amount to USD 134 billion per year (EUR 99 billion) if a common DALY value of EUR 45 000 per life year is applied (Jantunen et al., 2011). The impact of outdoor air pollution, which also influences the indoor environment, has been estimated even higher (Box 4.4).

The risks can, and should, be carefully managed. Following safe retrofitting procedures and taking a whole-building approach to energy efficiency interventions can avoid the potential pitfalls. Using safe building materials and adhering to indoor health protocols for adequate ventilation when undertaking refurbishment will minimise the potential for harmful indoor emissions. Additional safeguards should be put in place to manage the risk of renovation and construction activities releasing toxic particles (such as asbestos, lead paint and other carcinogens) into the indoor environment. Chemical, biological and physical pollution indoors can be contained at source; if this does not suffice, adequate ventilation will dilute the burden of pollution.

The US Department of Energy WAP provides an example of good practice: measuring for radon is a standard element in interventions, which also include installation of

8 Negative impacts on health and well-being may also arise from increasing noise caused by the use of heat pumps and ventilation systems. These can include annoyance, sleep disturbance, cardiovascular disease, impaired learning and memory, and higher stress levels (EEA, 2010).

9 DALY is an indicator drawn from the field of epidemiology.

© OECD/IEA, 2014.

carbon dioxide monitors and smoke detectors and maintenance of existing ventilation systems. A "do no harm" principle is effectively applied in the United States and Canada to ensure a focus on good building design and retrofitting while avoiding the risk of inadvertently inducing negative health impacts (US EPA, 2011; CELA, 2011). Most countries have introduced requirements to ensure minimum levels of ventilation in buildings and several best-practice guides to safe retrofitting procedures exist;[10] such policy tools have yet to be issued in all countries. Careful policy design, combined with adequate regulation and enforcement regimes, can strike a balance between good indoor air quality and the rational use of energy in buildings, while also avoiding the pitfalls of introducing energy efficiency measures into the complex system that buildings represent.

Box 4.4 Air quality in the outdoor environment and health impacts

Outdoor air quality is another major driver of health impacts. While in-depth analysis is beyond the scope of this chapter, the increasing importance of energy efficiency as a tool for combating local pollution warrants a special mention for its potential to deliver economic, environmental and health benefits.

Outdoor air pollution is estimated to cause 3.3 million premature deaths every year (Lim et al., 2012). Poorly planned and inefficient modes of transport are a major contributor; electricity generation technologies and energy consumption by buildings also play large roles. The problem is particularly acute in China, with the rapid increase in the number of motor vehicles and largely coal-based electricity generation. The World Bank estimates the total health cost associated with outdoor air pollution in urban areas of China in 2003 was between USD 25.2 billion and USD 83.5 billion (CNY 157 billion and CNY 520 billion), accounting for 1.2% to 3.3% of China's GDP (World Bank and SEPA, P. R. China, 2007).

Energy efficiency in the transport sector, through a combination of active travel*, public transport planning and lower-emission motor vehicles can play a major role in reducing local pollution (Dora, 2013; Macmillan et al., 2014; Rabl and Nazelle, 2011). Energy efficiency measures implemented in London transport systems are expected to save 7 300 DALYs per million people by 2030. The corresponding result for Delhi was almost 13 000 DALYs by 2030 (Woodcock et al., 2009).

Improving building insulation also has measurable impacts on outdoor concentrations of regional air pollutants. Very low-energy buildings were found to deliver emissions reduction of 9% in particulate matter and 6.3% for sulphur dioxide in northwestern Europe (Korsholm et al., 2012).

The role of energy efficiency in reducing air outdoor pollution is an area where uncertainties persist and further research is needed. Work is progressing quickly, however, led in particular by the Chinese government, which is working in several major cities to better understand the impacts of poor air quality on citizen health and the public health cost implications.

* Active travel includes any mode of travel that relies on human physical energy as opposed to motorised and carbon-fuelled modes, e.g. walking, running and cycling.

Methodological approaches

The list of potential health and well-being benefits linked to energy efficiency improvements is extensive. A central controversy remains, however, regarding how these benefits can be measured and how to generate results robust enough to inform energy efficiency policy making. Key challenges associated with strengthening the case for the multiple benefits

10 Examples include those issued by US EPA (2011) and the Canadian Environmental Law Association (CELA, 2011). Safe retrofit guidelines are currently being drafted for Northern Ireland.

© OECD/IEA, 2014.

approach for energy efficiency and health and well-being include mechanisms to collect data, criteria for valuing benefits, and making use of estimation methods where data are limited.

- **Collecting data** through empirical study of completed energy efficiency interventions is key to building up the evidence base in this area. Data needs can be met through methods such as in-depth intervention studies and cross-sectional surveys that look for associations across similar groups.
- Attributing **quantified values** to observed changes in health is vital to assessment. Various methods have been used, which can be fed into a chosen assessment methodology (see the Companion Guide at the end of this publication for more detailed discussion). Outcomes can be quantified either directly or through proxy indicators of change. Some evaluators have taken on the challenge of monetising health and well-being impacts, often by integrating value estimations into standard approaches such as BCA. The complex nature of health and well-being impacts means that qualitative assessment is important for capturing the subjective, experiential aspects of health.
- **Estimation methods** can generate prospective values to support *ex ante* assessment or provide alternatives when limited resources constrain empirical research. Modelling results will improve as more empirical data are generated to feed them; meanwhile, all estimation methods, including shortcut tools such as the use of multipliers, must be treated with caution.

Assessments should also consider temporal issues in project planning and evaluation design, taking into account both the duration of impacts from specific energy efficiency measures and the timeframe within which health and well-being impacts can be expected to emerge. Some health outcomes, such as asthma or mental health impacts, may be measureable in a relatively short time frame; many others can take several years to emerge.

Data challenges

By bridging past disciplinary boundaries, evaluators have transferred some unique and well-designed tools from epidemiology and public health to gather information about the impact of energy efficiency measures on health and well-being.

Community field trials are the most important study type for identifying whether an energy efficiency intervention makes a difference to programme recipients. After the first step of establishing baseline measurements has been completed, before and after measurements of indicators for exposure factors and for health and well-being can be taken, making it possible to quantify the changes associated with the intervention (Howden-Chapman et al., 2007, 2008; Barton, Basham and Shaw, 2004). This empirical data will then inform the development of prospective modelling efforts.

Less resource-intensive options include use of observational approaches such as cross-sectional surveys to draw inferences of cause and effect across like groups of people. The World Health Organization Large Analysis and Review of European housing and health Status project used a cross-sectional approach, correlating housing conditions with self-reported health status. Known limitations of this approach were minimised as the survey covered eight European cities, involving 3 373 dwellings and 8 519 individuals (Ormandy, 2009). Retrospective studies can also be carried out after an intervention, for example, by matching property records with anonymised national health records and comparing results measured in households having received interventions against a comparable group that did not receive any intervention (Telfar et al., 2011).

© OECD/IEA, 2014.

Most quantitative studies use standardised surveys to collect data, either drawn directly from the health or psychology profession[11] or developed by researchers on the basis of what is already known or hypothesised about the exposure-outcome pathway from energy efficiency to health. These provide a reliable basis for measuring the specific impacts on study participants and have proved particularly useful in measuring the effect of energy efficiency on mental health and well-being status (Liddell, 2013a; Gilbertson, 2013).

Quantifying outcomes

Evaluators have defined a range of indicators that can be used to track the impacts (both positive and negative) of improved indoor environments and lowered energy costs on physical and mental health and well-being. These indicators, the suitability of which depends on the case, facilitate quantification of health and well-being impacts so that they can be considered alongside values that are more familiar to energy efficiency experts and policy makers (i.e. measured energy cost savings).

As a starting point, an exposure approach can help to build understanding of when and how certain energy efficiency interventions generate conditions for positive health outcomes. Quantification of impacts can be used to inform design of future energy efficiency interventions and to build a body of supporting evidence for more detailed assessments of specific health outcomes (Box 4.5).

Monetising health and well-being impacts

Most energy efficiency policy assessments focus on comparing, in financial terms, implementation costs against the energy benefits generated. Translating quantified health and well-being impacts into monetary values substantially boosts their usefulness. By overlaying financial metrics onto identified health and well-being indicators, it is possible to derive the market values needed for current impact assessment approaches, such as calculation of a payback period, benefit-cost ratios or more comprehensive approaches (see the Companion Guide at the end of this publication).

The financial value of any energy cost savings provides a simple and direct measure for energy bill-related health outcomes. Another approach might be to calculate the amount of household income made available for spending on other health-supporting purchases such as medical advice and treatment to prevent illness, or the purchase of more nutritious food.

Beyond this, several concepts have been borrowed from epidemiology to provide metrics for monetising the health impacts of energy efficiency improvements. An important method is the use of "statistical life years", which can be calculated to represent the impact of a given illness and attributed a nominal financial value depending on country context.

- **Statistical life years** is a calculation used to measure the burden of a particular disease – or the value of a disease intervention – by the number of years of life a person is likely to lose from the disease or gain from the intervention. It can be applied to quantify the impact of any health impairment and also of death. The monetisation aspect is derived by assessing the value of an average human life, based on one person's earning power over a lifetime. This approach underlies a large proportion of total health impacts measured in studies to date: statistical life years gained tend to represent about 75% of health benefits in Europe and as much as 87% in some studies (Grimes et al., 2011). The measure is often translated into QALYs to take into account the quality of the life lived or DALYs to account for illnesses that lead to disability which affects quality of life. Care is needed in using this

11 The most common and reliable tests include the Health Survey (SF-36), the Warwick-Edinburgh Mental Well-being Scale, the Hospital Anxiety and Depression Scale, and the General Health Questionnaire 12.

© OECD/IEA, 2014.

concept, however, as it is impossible to place an entirely objective financial value on life and the concept will be approached very differently across countries.[12]

Innovative studies have identified other proxy measures for monetising health and other outcomes that can be found in the flow-on impacts of improved health.

Box 4.5 Quantified indicators for impacts arising from energy efficiency improvements

To address the challenge of drawing causal connections between energy efficiency policy and potential health outcomes, it is helpful to start by using exposure indicators to track changes in the indoor environment and in energy expenditure. Once sufficient causal evidence has been found in exposure, a further set of indicators, relevant to the anticipated health and well-being outcomes, can be considered.

To establish the links between intervention, exposure and impacts, evaluators need to make assumptions about the effect of identified housing conditions on health and well-being. Such assumptions should be based on knowledge from health and social disciplines. To derive the results outlined in this chapter, experts have identified a range of quantified indicators for health outcomes. Some of the most common indicators used to quantify health and well-being impact studies are listed in Table 4.2.

Table 4.2 Common indicators used in measuring health and well-being impacts of energy efficiency

Exposure indicators	Health and well-being indicators
Average indoor temperature of building	Number of statistical life years lost (see below)
Humidity levels inside the building	Rate of excess seasonal mortality
Number of rooms heated and in active use by occupants	Number of visits to the hospital, doctor, pharmacist
Level of particular indoor pollutants	Cost of treatment (see below)
Number or percentage of buildings using harmful fuels for heating	Number of accidents or injuries within the building
Number or percentage of occupants reporting improved indoor comfort	Assessed health status (based on existing standardised surveys or self-assessment)
Percentage of monthly budget spent on energy bills	Number of restricted activity days (number of days off work or school)

Both sets of indicators should be measured before and after an intervention, with the underlying aim of identifying the circumstances in which an energy efficiency intervention that reduces the exposure risk has a positive impact on the health and well-being indicator. While many health and well-being indicators have a strong qualitative element, others are quite objective and can be measured for inclusion in traditional quantitative assessment methods.

Social science has developed an approach called the *Exposome*, which draws on trans-disciplinary methods to assess the effects of cumulative lifelong environmental exposures on public health arising from four environmental domains: the natural; the built; the social; and the public policy domain.*

* http://communitymappingforhealthequity.org/public-health-exposome-data/.

12 Annual rates for one QALY have been set, for example, in the United Kingdom at GBP 30 000 (USD 50 450), in the European Union at EUR 45 000 (USD 61 650) and in New Zealand at NZD 30 000 (USD 25 775).

© OECD/IEA, 2014.

- **Cost of treatment** calculates the financial outlay associated with health problems, or indeed the costs avoided when health and well-being improves. The costs for illness, disease and discomfort, for example, might include the price of a doctor's visit, one night in hospital, pharmaceuticals and other medical costs. These costs can be borne or avoided by the individual (e.g. the purchase price of the health-related goods and services) or by the state or society (e.g. the cost of state health services and subsidies). The perspective from which these benefits should be considered will depend on the entity bearing the costs against which they are being compared – i.e. whether the energy efficiency intervention was publically or privately funded.
- **Days off school** can be translated into monetary terms by overlaying the value of lost earnings for the parent or the cost of hiring a caregiver for the sick child. Loss of future earnings can also be used to monetise the impact of reduced educational attainment for the child through days off school (Box 4.6).
- **Days off work** and associated productivity impacts can be monetised through a calculation of lost earnings (Howden-Chapman et al., 2009). The broader social impact of this in terms of reduced productivity has been assessed for an entire society.

Box 4.6 Using reduced absenteeism to calculate the financial value of improved health

Concrete evidence exists of the flow-on effect that healthy homes have in reducing absenteeism from school and work.

The causal link has been drawn most strongly for energy efficiency retrofit programmes that improve respiratory health of child occupants, leading to a reduced number of school absences (Howden-Chapman et al., 2008; 2009; 2012; Preval et al., 2010). A 15% reduction in days off school has been measured among children in homes that received energy efficiency upgrades (Free et al., 2010).

Fewer days off work have been similarly reported as a result of adult health improvements, with an odds ratio of 0.6 of participants having had days off work due to illness (Chapman et al., 2009; Howden-Chapman et al., 2007; Laing and Baker, 2006).

The impact of time off school has been monetised through the daily cost of hiring a caregiver for sick children (six hours employment at the minimum wage rate); the impact for time off work is assessed through lost earnings. Including the value of reduced absenteeism has been shown to increase the NPV of overall benefits by 11.5% (Chapman et al., 2009). More work in more regions will strengthen these initial assumptions and findings.

Qualitative approaches

A requirement for monetisation clashes with the complexity of health as a product of many interactions of physical, economic, social and cultural factors, which are strongly subjective in their nature. Much of the evidence available to date is based on self-reported health improvements. The quality of the evidence varies with the sample sizes and robustness of methods used, with data remaining largely qualitative.

For many health and well-being impacts, a direct equivalent in monetary terms has not yet been determined. In some cases, monetisation may not be appropriate, but often, the challenge might simply be a lack of expertise needed to carry out the quantification of certain impacts – for example, in cases where energy experts responsible for carrying out assessments lack access to data and input from health and well-being experts. Where quantification is not practical, experts advocate for the triangulation of evidence from

© OECD/IEA, 2014.

mixed methodologies to ensure that assessments take into account all impacts reported by recipients of an energy efficiency intervention.

Qualitative evidence is generally gathered from case studies, focus groups, systemic interviews and surveys that ask programme participants to describe the impacts they experienced (see the Companion Guide at the end of this publication). In the health and well-being context, qualitative research has been used to explore occupant reactions to retrofitted insulation and the ways in which they experience and cope with fuel poverty (Critchley et al., 2007; O'Sullivan, Howden-Chapman and Fougere, 2011). In-depth interviews of participants in a retrofit programme show that householders have more control over their home environment, with a reported impact on physical and mental well-being as well as improved ability to self-manage long-term conditions (Bashir et al., 2014). Although this evaluation was not able to quantify impacts, case studies provided powerful evidence of how the programme helped to promote social connections for householders who were previously isolated. Methods exist to build on subjective descriptions, for example by asking respondents to value the impacts as a fraction of bill savings or some other observable value (an example of contingent valuation) (Tetra Tech, Inc. and Massachusetts Program Administrators, 2011).

Assessment approaches need to accommodate qualitative data if they are to more fully reflect health and well-being outcomes. Holistic assessment frameworks, such as multi-criteria decision analysis (MCDA) which combines qualitative evidence of costs and benefits with quantified benefit-cost assessments, are important tools in the health and well-being context. These and other options are described in more detail in the Companion Guide at the end of this publication.

Estimation methods

The empirical investigation of actual energy efficiency interventions described above is the most robust way to gain an understanding of the complex interactions at work in generating health and well-being impacts. In practice, time and resource limitations make it difficult to obtain such robust results. As an alternative, estimations based on quantified results from previous studies provide an extremely valuable input into the policy-making process, particularly in *ex ante* policy assessment. Several approaches have been used in estimating health impacts, including health impact assessments (HIAs) based on data about exposure factors, detailed modelling efforts and a range of simpler (but much less reliable) shortcut methods.

HIAs facilitate assessment of the potential health impacts of policies and plans in both qualitative and quantitative terms. These are systematically applied in the context of *ex ante* appraisal of policy proposals to give a preliminary estimate of expected impacts. Several tools have been developed that can support the HIA process by measuring the risk posed to health by various exposure factors in buildings. Examples include the *British Housing Health and Safety Rating System* (UK DCLG, 2006; Davidson et al., 2010), the New Zealand *Healthy Homes Index* (Gillespie-Bennett et al., 2013; McClean Salls et al., 2013) and the United States *National Healthy Housing Standard* (US NCHH, 2013).

Quantitative **HIA models** are used to generate detailed quantified estimates of effects of energy efficiency on health to support *ex ante* policy evaluation. Models use data from existing empirical studies and additional simulation work to derive assumptions that can support policy making. Two recent examples include the Health Impacts of Domestic Energy Efficiency Measures (HIDEEM) model recently developed in the United Kingdom (Box 4.7) and the Net Benefit Model in New Zealand (see the Companion Guide at the end of this publication).

© OECD/IEA, 2014.

Box 4.7 Modelling the health impacts of energy efficiency measures

Recognising the need for more robust evaluation, experts at the London School of Hygiene and Tropical Medicine and University College London developed the Health Impacts of Domestic Energy Efficiency Measures (HIDEEM) model. The model aims to do five things: provide estimates of indoor exposure to cold and air pollutants in UK housing stock; define a set of residential energy efficiency measures; measure the resulting changes in exposures and consequential health impacts; measure the resulting changes in energy demand; and monetise the estimated health impacts.

The health impacts from changes in exposure due to energy efficiency are applied to a representative population based on the latest English Housing Survey, taking into account the disease risk of the population. The model applies selected efficiency measures to target dwelling or household groups, and then produces estimates of energy cost saving and of health impacts (morbidity and mortality) in QALYs (monetised at USD 34 000 to USD 51 000 per QALY). For economic analysis, energy efficiency interventions are costed along with health care costs/savings. The model can be used to undertake either CEA or BCA.

The cost-effectiveness ratios shown below are the "costs per QALY gained" when applying a suite of energy efficiency and ventilation measures to low-income English households (~4 million) and examining cold-related health impacts. A "government and household perspective" is taken that includes the costs of the intervention, energy cost saving and health care costs. The results suggest that loft and cavity wall insulation produce a net cost saving, as the savings in energy and healthcare costs outweigh those of the intervention (Figure 3.1). For solid wall insulation, double glazing and boiler replacement (with a condensing boiler) there is a greater cost per QALY achieved, and so too for the combined intervention. For the overall combined set of measures, the approximate results in costs or QALYs over 42 years (the average lifetime of the installed energy efficiency measures) are:

- USD 10 885 (GBP 6 410) per capita in intervention costs (USD 255 [GBP 150] per person, per year)
- USD 8 540 (GBP 5 030) per capita in energy demand reduction (USD 204 [GBP 120] per person, per year)
- USD 68 (GBP 40) per capita in healthcare cost savings (USD 0.40 [GBP 0.24] per person, per year)
- 210 QALYs gained per 10 000 persons (USD 492 [GBP 290] gained per capita; USD 12 [GBP 7] per person, per year).

Figure 4.1 HIDEEM results for cost-effectiveness of energy efficiency for low-income households

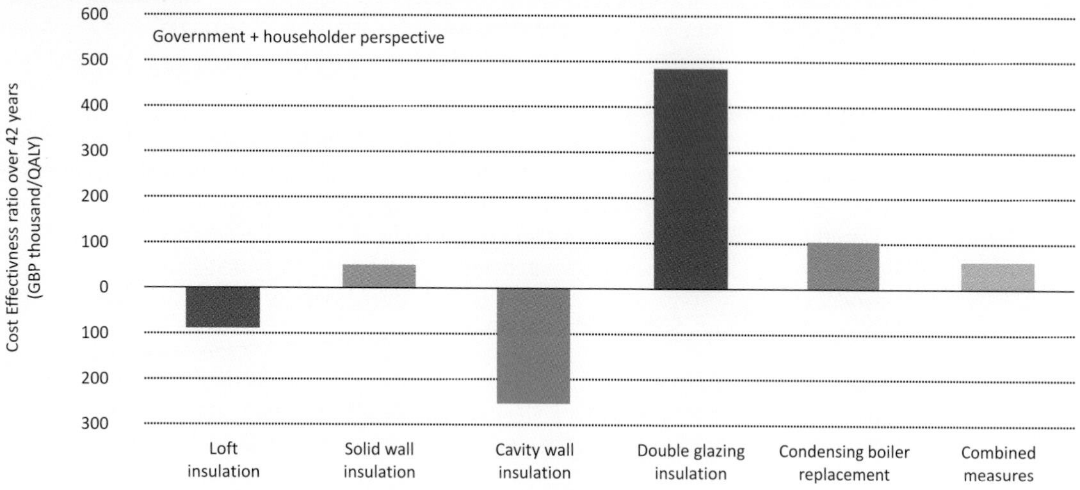

Government + householder perspective
Cost Effectiveness ratio over 42 years (GBP thousand/QALY)
600
500
400
300
200
100
0
100
200
300
Loft insulation
Solid wall insulation
Cavity wall insulation
Double glazing insulation
Condensing boiler replacement
Combined measures

© OECD/IEA, 2014.

A long-term aim of the UK Department of Energy and Climate Change is to incorporate evidence from this project into mainstream policy development and assessment processes, so that policies can be designed to maximise health benefits and reduce potential unintended consequences (LSHTM and UCL, 2013).

Note: All figures in this box converted from GBP. Only cold-related health impacts are included in these calculations. Inclusion of impacts related to air pollutants which would boost these figures significantly, are excluded because results are undergoing revisions.

Development of estimation tools requires a significant investment of research and funding, and models need to be regularly updated with new information gathered from empirical work. Continued effort to build the empirical evidence base will enhance the ability of the health impact models to predict the complex effects of energy efficiency interventions on health and well-being. A broad range of assumptions is necessary in all modelling of this type, so it is important to note the inherent uncertainty of model outputs and to test them using sensitivity analysis.

Where limited evaluation of health impacts has been done locally, **desk-top comparison** is a useful, low-resource estimation approach. Essentially, the method draws on results of previous studies to compare and explain different outcomes recorded in different countries. **Statistical regression analysis** can be used to isolate the various causes of disparities in health outcomes between studies, and to draw approximate conclusions about the impacts of energy efficiency for use in a high-level decision-making context. As a shortcut, using **default values** from existing studies might offer a first-pass estimate of the range of potential health impacts of a proposed energy efficiency policy. Many variables can influence the costs and health benefits of a given energy efficiency intervention; given the lack of consensus on methods in this area, direct transferability of results is currently limited (see the Companion Guide at the end of this publication).

Policy-making considerations

The health and well-being impacts described above can all be triggered by building-related exposure factors that are closely linked to energy efficiency. Their occurrence is influenced by numerous mediating factors including: the baseline conditions of a building; the health, housing situation and income of the occupants; the level of energy efficiency and the nature, scale and delivery mode of interventions; any trade-off made through the behaviour choices of the occupants; and the cost of fuels. Some of these factors can be influenced by policy design; thus, it is important to consider how policy choices drive health and well-being impacts. Three key factors warrant particular attention: targeting of energy efficiency measures, the scale of interventions, and efforts to co-ordinate action on energy and health.

Targeting interventions

The characteristics of individuals receiving energy efficiency interventions strongly influence the extent of health outcomes that can be expected. Improvements in health are most likely when measures target those living in poor housing (particularly in conditions of inadequate warmth) or with pre-existing illnesses (Thomson et al., 2013). Targeting can be done in several ways, the two most common being: i) targeting occupants (e.g. based on receipt of benefits, lifecycle stage or medical risk, or fuel-poor status); or ii) targeting dwellings, based on their level of efficiency.

Examples of programmes that target the fuel-poor and other at-risk groups (the elderly, people with disabilities and young children) include the US Weatherization Assistance Program (WAP), the French national fuel poverty programme *Habiter Mieux* (Live Better)

© OECD/IEA, 2014.

and the Affordable Warmth Programme in Northern Ireland (Walker et al., 2013). The last two have a particular focus on the elderly and proactively seek to target households in fuel poverty. Targeting these groups is an effective means of ensuring that energy efficiency interventions yield the greatest benefits and can increase the value of outcomes by 45%[13] (Grimes et al., 2011). More accurate targeting requires a longer pre-implementation process to identify the appropriate recipients of a programme and implies a greater administrative burden. The additional burden of more accurate targeting needs to be balanced against the increased benefits generated when energy efficiency measures reach households most in need.

Programmes implemented in England and New Zealand have had success with targeting dwellings. This approach has the advantage of avoiding the uncertainty and cost of identifying individuals most in need, while being sure to address the least efficient dwellings.

More recent approaches have attempted targeting on the basis of a "combined index" using housing characteristics (e.g. poor energy efficiency, age of house, etc.) and demographic data about the people living in them (Walker et al., 2013). This targeting system, based on multiple risks, has delivered 78% accuracy in reaching at-risk populations (Liddell and McKenzie, 2013). Combining broad approaches with bottom-up identification of vulnerable households in a more personalised way – for example, by leveraging healthcare professionals' relationships with their patients – can further refine identification of the most vulnerable and prove cost-effective.

Issues of equity may come into play in targeting, where measures are offered only to certain sections of the population; this must be balanced against the potential for improving health equality for vulnerable populations. The most appropriate balance is likely to be country-specific; this and other issues that require careful consideration of the country context are explored further in the Companion Guide at the end of this publication.

Scale of interventions

The relative value of comprehensive retrofits as opposed to smaller interventions has been underlined by experience in several countries. While insulation measures have the greatest individual impact on health and well-being, a more extensive retrofit package (e.g. including insulation, heating measures, draught-proofing, ventilation and more) can be expected to deliver greater benefits while also saving on implementation costs. Small additional measures, such as installation of smoke and radon detectors, can make a large contribution to retrofit outcomes. For example, external features such replacing windows or even the front door can enhance both the actual and the perceived benefits for recipients (Liddell and Guiney, 2014). In the United States, a "one touch" approach has been used to deliver multiple benefits for health, housing (including safety) and energy combined into one holistic intervention, with the advantage of potentially minimising administrative costs as well as disruption to the recipient household.

Co-ordinated action at the intersection of energy and health

Health and well-being crosses disciplinary boundaries. This has important implications for energy efficiency agencies, but also those dealing with health, housing, social welfare, urban planning and, to the extent that issues related to air pollution come into play, those dealing with environment, transport and industry.

13 This intervention saw greatest positive health and well-being impacts among participants in fuel poverty, with existing illness.

© OECD/IEA, 2014.

In the United States, innovative resource sharing among government departments has helped to distribute the resource burden of thorough assessment of multiple benefits of its WAP. Specifically, the US Environmental Protection Agency (US EPA) contributes funding and scientific analysis services to the WAP research team in return for the installation of US EPA radon measurement devices as part of weatherisation measures. Local community action agencies are often tasked with administering federal level programmes for the fuel poor, or energy utilities are required to run energy efficiency programmes to support their lowest income customers. Inter-agency and interdisciplinary co-operation (Chapter 7) supports a more comprehensive understanding of the complex interactions at work in the multiple benefits approach. It also helps to identify new sources for data that could support analysis of health and well-being impacts, and ultimately to design policies that address shared national objectives.

Further research for stakeholders

To facilitate more systematic integration of health and well-being impacts in energy efficiency policy, efforts to build coherence in methodological approaches and measurement scales should be scaled up to build on the substantial work already underway (Table 4.3).

Table 4.3 Further stakeholder research and collaboration opportunities in health and well-being

Area	Specific actions
Benefit areas and causal linkages	Investigate how improved energy efficiency in other sectors (i.e. buildings, transport or energy generation) might influence health and well-being.
	Investigate the health and well-being impacts of reduced outdoor air pollution through improved energy efficiency.
	Investigate the social outcomes of neighbourhood-wide retrofitting programmes (e.g. improved social cohesion and hypothesised outcomes such as a contribution to lower crime rates).
	Build evidence on potential mental health impacts of energy efficiency.
Methodologies and approaches	Advance comparative studies on the benefits of targeting houses or occupants.
	Explore trans-disciplinary approaches to understanding flow-on social impacts.
	Refine the various methodological approaches to support standardisation.
Data, indicators and metrics	Resolve the construct validity of different measurement scales for mental health used in the context of energy efficiency impacts.
	Strengthen methods and assumptions to quantify the effect of reduced absenteeism (e.g. in different regions).
Collaborative initiatives	Develop an international database for health and well-being impacts of energy efficiency programmes.

Conclusions

The significant health and well-being benefits that have been recorded following energy efficiency improvements present a compelling argument for increased efforts to understand what energy efficiency can deliver directly and indirectly to individuals and communities, and from a public health perspective.

© OECD/IEA, 2014.

The evidence base for direct impacts on physical health is growing, and a case is also emerging for impacts on mental health and a range of more indirect impacts. Much work is still needed to better understand the exposure-outcome pathway between the impacts of energy efficiency measures on indoor environments and energy bills, and the health and well-being outcomes they can generate. Increasingly robust measurement approaches are being applied and useful indicators are being developed to enable impacts to be quantified through use of health sector data and proxies found further afield. Because of the highly subjective nature of self-reported health and well-being status, rigorous assessment using mixed methodologies will remain essential for capturing the complexities of health and well-being impacts. Ultimately, assessment frameworks may need to evolve in order to accommodate this reality.

The growing evidence base has facilitated the development of models for estimating health and well-being impacts. In turn, these models could support the inclusion of health and well-being impacts as stated objectives within new energy efficiency measures, thereby supplying the policy decision-making process with increasingly robust estimations of potential policy impacts.

Bibliography

ACE (Association for the Conservation of Energy) (2014), *The Energy Bill Revolution. Fuel Poverty: 2014 Update*, ACE, London, www.e3g.org/docs/ACE_and_EBR_fact_file_%282014-02%29_Fuel_Poverty_update_2014.pdf (accessed 13 May 2014).

Anderson, D.R. et al. (2000), "The relationship between modifiable health risks and group level health care expenditures", *American Journal of Health Promotion*, Vol. 15, No. 1, AJHP (American Journal of Health Promotion) and Allen Press, North Hollywood, pp. 45-52, www.ncbi.nlm.nih.gov/pubmed/11184118 (accessed 24 June 2014).

Barton, A., M. Basham and S. Shaw on behalf of Torby Healthy Housing (THH) Group (2004), *Central Heating: Uncovering the Impact on Social Relationships in Household Management*, final report to the EAGA Partnership Charitable Trust, Peninsula Medical School, Plymouth www.eagacharitabletrust.org/index.php/projects/item/central-heating-uncovering-the-impact-on-social-relationships-and-household-management (accessed 20 June 2014).

Bashir N. et al. (2014), *An Evaluation of the FILT Warm Homes Service CRESR*, Sheffield Hallam University, Sheffield, www.shu.ac.uk/research/cresr/sites/shu.ac.uk/files/eval-filt-warm-homes.pdf (accessed May 2014).

Bird J., R. Campbell and K. Lawton (2010), *The Long Cold Winter: Beating Fuel Poverty*, IPPR (Institute for Public Policy Research) and NEA (National Energy Action), London, www.energy-uk.org.uk/publication/finish/6/284.html (accessed 23 June 2014).

Boardman, B. (2010), *Fixing Fuel Poverty: Challenges and Solutions*, Earthscan, London.

Bone, A. et al. (2010), "Will drivers for home energy efficiency harm occupant health?", *Perspectives in Public Health*, Vol. 130, No. 5, Royal Society for Public Health, Los Angeles, pp. 233-238, http://rsh.sagepub.com/content/130/5/233.long (accessed 30 June 2014).

Braubach, M. and A. Ferrand (2013), "Energy efficiency, housing, equity and health", *International Journal of Public Health*, Vol. 58, No. 3, Springer Birkhäuser, Basel, pp. 331-332, http://download.springer.com/static/pdf/149/art%253A10.1007%252Fs00038-012-0441-2.pdf?auth66=1403830690_d59e638d43621160d630e5ecf80ac743&ext=.pdf (accessed 24 June 2014).

Campion J. et al. (2012), "European Psychiatric Association (EGA) guidance on prevention of mental disorders", *European Psychiatry*, Vol. 27, No. 2, Editions scientifiques Elsevier, Paris, pp. 68-80, www.ncbi.nlm.nih.gov/pubmed/22285092 (accessed 25 June 2014).

CELA (Canadian Environmental Law Association) (2011), *Healthy Retrofits: The Case for Better Integration of Children's Environmental Health Protection into Energy Efficiency Programs*, CELA, Toronto, www.cela.ca/publications/healthy-retrofits-full-report (accessed November 2013).

© OECD/IEA, 2014.

Chapman, R. et al. (2009), "Retrofitting housing with insulation: A cost benefit analysis of a randomised community trial", *Journal of Epidemiology and Community Health*, Vol. 63, No. 4, BMJ Publishing Group Ltd., London, pp. 271-277, www.ncbi.nlm.nih.gov/pubmed/19299400 (accessed 25 June 2014).

Christidis N., G.C. Donaldson and P.A. Stott (2010), "Causes for the recent changes in cold- and heat-related mortality in England and Wales", *Climatic Change*, Vol. 102, No. 3-4, Springer Netherlands, Houten, pp. 539-553, http://link.springer.com/article/10.1007%2Fs10584-009-9774-0#page-1 (accessed 25 June 2014).

Clare, S. (2013), "Feeling cold: Phenomenology, spatiality, and the politics of sensation", *Differences*, Vol. 24, No. 1, Duke University Press, Durham, pp. 169-191, http://differences.dukejournals.org/content/24/1/169.abstract (accessed 25 June 2014).

Clinch, J.P. and J.D. Healy (2001), "Cost-benefit analysis of domestic energy efficiency", *Energy Policy*, Vol. 29, No. 2, Elsevier Ltd., Amsterdam, pp. 113-124, www.sciencedirect.com/science/article/pii/S0301421500001105 (accessed 25 June 2014).

Clinch, J.P. and J.D. Healy (2000), "Housing standards and excess winter mortality", *Journal of Epidemiology and Community Health*, Vol. 54, No. 9, BMJ Publishing Group Ltd., London, pp. 719-720, www.ncbi.nlm.nih.gov/pmc/articles/PMC1731747/pdf/v054p00719.pdf (accessed 25 June 2014).

Cole, B.L. and J.E. Fielding (2007), "Health impact assessment: A tool to help policy makers understand health beyond health care", *Annual Review of Public Health*, Vol. 28, No. 1, Annual Reviews, Palo Alto, pp. 393-412, ftp://prmdftp.sonoma-county.org/Health%20Impact%20Assessment%20Training%20Materials/Required%20Readings/Day1_2.%20Cole.ARPH.pdf (accessed 25 June 2014).

Cook, J.T. et al. (2008), "A brief indicator of household energy security: Associations with food security, child health, and child Development in US infants and toddlers", *Pediatrics*, Vol. 122, No. 4, American Academy of Pediatrics, Elk Grove Village, pp. e867-e875, www.ncbi.nlm.nih.gov/pubmed/18829785 (accessed 25 June 2014).

Copenhagen Economics (2012), *Multiple Benefits of Investing in Energy-efficient Renovation of Buildings – Impact on Finances*, report commissioned by Renovate Europe, Renovate Europe, Brussels, www.renovate-europe.eu/uploads/Multiple%20benefits%20of%20EE%20renovations%20in%20buildings%20-%20Full%20report%20and%20appendix.pdf (accessed 23 June 2014).

Critchley, R. et al. (2007), "Living in cold homes after heating improvements: Evidence from Warm-Front, England's Home Energy Efficiency Scheme", *Applied Energy*, Vol. 84, No. 2, Elsevier Ltd., Amsterdam, pp. 147-158, www.sciencedirect.com/science/article/pii/S0306261906000791 (accessed 25 June 2014).

Davidson, M. et al. (2010), *The Real Cost of Poor Housing*, IHS BRE (Building Research Establishment) Press, Watford, www.brebookshop.com/details.jsp?id=325401 (accessed 25 June 2014).

Davies, M. and T. Oreszczyn (2012), "The unintended consequences of decarbonising the built environment: A UK case study", *Energy and Buildings*, Vol. 46, Elsevier B.V., Amsterdam, pp. 80-85, www.sciencedirect.com/science/article/pii/S0378778811005068 (accessed 25 June 2014).

De Haro, M.T. and A. Koslowski (2013), "Fuel poverty and high-rise living: using community-based interviewers to investigate tenants' inability to keep warm in their homes", *Journal of Poverty and Social Justice*, Vol. 21, No. 2, Policy Press, Bristol, pp. 109-121, www.social-policy.org.uk/lincoln2012/Koslowski%20P7.pdf (accessed 25 June 2014).

Delussu, S. (2014), *Evaluation of the Effects on Cardiovascular and Metabolic Systems of at Work Stair-climbing Intervention Program for Employees*, unpublished report for the Italian government.

Dempsey, N. et al. (2011), "The social dimension of sustainable development: Defining urban social sustainability", *Sustainable Development*, Vol. 19, No. 5, John Wiley & Sons, Ltd. and ERP Environment, Hoboken, pp. 289-300, http://onlinelibrary.wiley.com/doi/10.1002/sd.417/pdf (accessed 25 June 2014).

Dora, C. (2013), "Energy is a health issue: The WHO approach to evaluating health benefits of a green economy", presentation at the IEA-EEA Roundtable on the Health & Well-being Impacts of Energy Efficiency Improvements, Copenhagen, 18-19 April 2013, www.iea.org/workshop/roundtableonthehealthwell-beingimpactsofenergyefficiencyimprovements.html (accessed 22 June 2014).

Dubois, U. (2013), "Identifying fuel poor households to maximize the benefits of energy efficiency interventions", presentation at the IEA-EEA Roundtable on the Health & Well-being Impacts

© OECD/IEA, 2014.

of Energy Efficiency Improvements, Copenhagen, 18-19 April 2013, www.iea.org/workshop/roundtableonthehealthwell-beingimpactsofenergyefficiencyimprovements.html (accessed 22 June 2014).

Ebi, K.L. and D. Mills (2013), "Winter mortality in a warming climate: a reassessment", *Wiley Interdisciplinary Reviews: Climate Change*, Vol. 4, No. 3, WIREs Climate Change, John Wiley & Sons Ltd. in association with the Royal Meteorological Society and the Royal Geographical Society (with IBG), Hoboken, pp. 203-212, http://wires.wiley.com/WileyCDA/WiresArticle/wisId-WCC211.html (accessed 25 June 2014).

EC VCWG (European Commission Vulnerable Consumer Working Group) (2013), *Guidance Document on Vulnerable Consumers*, Report November 2013, EC, Brussels, http://ec.europa.eu/energy/gas_electricity/doc/forum_citizen_energy/20140106_vulnerable_consumer_report.pdf (accessed 13 May 2014).

ECA (European Collaborative Action on Urban Air, Indoor Environment and Human Exposure) (2013), *Report No. 30: Guidelines for Health-based Ventilation in Europe (HEALTHVENT)*, Technical University of Denmark, Lyngby, www.efanet.org/health-based-ventilation-guidelines-for-europe-healthvent (accessed 25 June 2014).

EEA (European Environment Agency) (2010), *Good Practice Guide on Noise Exposure and Potential Health Effects*, EEA Technical Report No 11/2010, EEA, Copenhagen, www.eea.europa.eu/publications/good-practice-guide-on-noise (accessed 25 June 2014).

Fisk, W.J., D. Black and G. Brunner (2011), "Benefits and costs of improved IEQ in U.S. offices", *Indoor Air*, Vol. 21, No. 5, Blackwell, Oxford, pp. 357–367, www.ncbi.nlm.nih.gov/pubmed/21470313 (accessed 28 June 2014).

Fisk, W. and O. Seppanen (2007), "Providing better indoor environmental quality brings economic benefits", *Proceedings of Clima 2007 Well Being Indoors*, Helsinki, 10-14 June 2007, FINVAC, Helsinki, http://energy.lbl.gov/ied/sfrb/pdfs/performance-1.pdf (accessed 21 June 2014).

Fisk, W. and O. Seppanen (2006), "Some quantitative relations between indoor environmental quality and work performance or health", *HVAC&R Research*, Vol. 12, No. 4, ASHRAE (American Society of Heating, Refrigerating, and Air-Conditioning Engineers), Inc., Atlanta, pp. 957-973, https://publications.lbl.gov/islandora/object/ir%3A151590/datastream/PDF/view (accessed 21 June 2014).

Frank, D.A. et al. (2006), "Heat or eat: The low income energy assistance program and nutritional and health risks among children less than 3 years of age", *Pediatrics*, Vol. 118, No. 5, American Academy of Pediatrics, Elk Grove Village, pp. e1293-e1302, http://pediatrics.aappublications.org/content/118/5/e1293.full.pdf (accessed 28 June 2014).

Free, S. et al. (2010), "Does more effective home heating reduce school absences for children with asthma?", *Journal of Epidemiology and Community Health*, Vol. 64, No. 5, BMJ Publishing Group Ltd., London, pp. 379-386, http://jech.bmj.com/content/64/5/379.full.pdf+html (accessed 28 June 2014).

Galiotto, N., P. Heiselberg and M. Knudstrup (2014), *The Integrated Renovation Process – A Holistic Methodology Towards Nearly Zero Energy Buildings*, Technical Report No. 165, Department of Civil Engineering, Aalborg University, Alalborg.

Gilbertson J. (2013), "Psychosocial routes from housing investment to health: Evidence from England's home energy efficiency scheme", presentation at the IEA-EEA Roundtable on the Health & Well-being Impacts of Energy Efficiency Improvements, Copenhagen, 18-19 April 2013, www.iea.org/workshop/roundtableonthehealthwell-beingimpactsofenergyefficiencyimprovements.html (accessed 22 June 2014).

Gilbertson, J., M. Grimsley and G. Green (2012), "Psychosocial routes from housing investment to health gain. Evidence from England's home energy efficiency scheme", for the Warm Front Study Group, *Energy Policy*, Vol. 49, Elsevier Ltd., Amsterdam, pp. 122-133, www.sciencedirect.com/science/article/pii/S0301421512000791 (accessed 28 June 2014).

Gillespie-Bennett, J. et al. (2013), "Improving our nation's health, safety and energy efficiency through measuring and applying basic housing standards: Viewpoint article", *New Zealand Medical Journal*, Vol. 126, No. 1279, New Zealand Medical Association, Wellington, p. 1379, http://journal.nzma.org.nz/journal/126-1379/5763/.

Greening L.A., D.L. Greene and C. Difiglio (2000), "Energy efficiency and consumption – the rebound effect – a survey", *Energy Policy*, Vol. 28, No. 6-7, Elsevier Ltd., Amsterdam, pp. 389-401, www.sciencedirect.com/science/article/pii/S0301421500000215 (accessed 28 June 2014).

© OECD/IEA, 2014.

Grimes A. et al. (2011), *Cost Benefit Analysis of the Warm Up New Zealand: Heat Smart Program*, report for the Ministry of Economic Development, Motu Economic and Public Policy Research, Wellington, www.motu.org.nz/files/docs/NZIF_CBA_report_Final_Revised_0612.pdf (revised June 2012, accessed 28 June 2014).

Grynszpan D. (2003), "Lessons from the French heatwave", *The Lancet*, Vol. 362, No. 9391, Elsevier Ltd., Amsterdam, pp. 1169-1170, www.thelancet.com/journals/lancet/article/PIIS0140-6736(03)14555-2/fulltext#article_upsell (accessed 28 June 2014).

Healy, J.D. (2003), "Excess winter mortality in Europe: a cross country analysis identifying key risk factors", *Journal of Epidemiology and Community Health*, Vol. 57, BMJ Publishing Group Ltd., London, pp. 784-789, http://jech.bmj.com/content/57/10/784.full.pdf+html (accessed 28 June 2014).

Hills, J. (2012), *Getting the Measure of Fuel Poverty: Final Report of the Fuel Poverty Review*, CASE Report 72, commissioned by DECC (Department of Energy and Climate Change), Crown copyright, CASE (Centre for the Analysis of Social Exclusion), London, http://sticerd.lse.ac.uk/dps/case/cr/CASEreport72.pdf (accessed 28 June 2014).

Hong S. H. et al. (2009), *A Field Study of Thermal Comfort in Low-income Dwellings in England before and after Energy Efficient Refurbishment*, http://discovery.ucl.ac.uk/15210/1/15210.pdf (accessed 3 July 2014).

HOPE (Health Optimisation Protocol for Energy-efficient Buildings) Project (2005), *Health Optimisation Protocol for Energy Efficient Buildings: Pre-normative and Socioeconomic Research to Create Healthy and Energy Efficient Buildings*, EU Energy, Environment and Sustainable Development, EC, Brussels, http://hope.epfl.ch/results/FinalReportHOPEpublic.pdf (accessed 21 June 2014).

Howden-Chapman, P. (2013), "Capturing the health benefits of energy efficiency programs in New Zealand – a success story", presentation at the IEA-EEA Roundtable on the Health & Well-being Impacts of Energy Efficiency Improvements, Copenhagen, 18-19 April 2013, www.iea.org/workshop/roundtableonthehealthwell-beingimpactsofenergyefficiencyimprovements.html (accessed 22 June 2014).

Howden-Chapman, P. et al. (2012), "Tackling cold housing and fuel poverty in New Zealand: A review of policies, research, and health impacts", *Energy Policy*, Vol. 49, Elsevier Ltd., Amsterdam, pp. 134-142, www.sciencedirect.com/science/article/pii/S0301421511007336 (accessed 28 June 2014).

Howden-Chapman, P. et al. (2009), "Warm homes: drivers of the demand for heating in the residential sector in New Zealand", *Energy Policy*, Vol. 37, No. 9, Elsevier Ltd., Amsterdam, pp. 3387–3399, www.sciencedirect.com/science/article/pii/S0301421508007647 (accessed 28 June 2014).

Howden-Chapman, P. et al. (2008), "Effects of improved home heating on asthma in community dwelling children: randomised community study", *British Medical Journal*, Vol. 337, British Medical Association, London, pp. 852-855, www.ncbi.nlm.nih.gov/pmc/articles/PMC2658826/ (accessed 28 June 2014).

Howden-Chapman, P. et al. (2007), "Retrofitting houses with insulation to reduce health inequalities: results of a clustered, randomised trial in a community setting", *British Medical Journal*, Vol. 334, British Medical Association, London, pp. 460-464, www.bmj.com/highwire/filestream/348503/field_highwire_article_pdf/0/460 (accessed 28 June 2014).

IEA (International Energy Agency) (2011), *Evaluating the Co-benefits of Low-income Energy-efficiency Programmes: Results of the Dublin Workshop 27-28 January 2011*, Workshop Report, OECD/IEA, Paris, www.healthyhousing.org.nz/wp-content/uploads/2012/08/low_income_energy_efficiency_Examining-the-CoBenefits2011.pdf (accessed 28 June 2014).

Jantunen M. et al. (2011), *Promoting Actions for Healthy Indoor Air (IAIAQ)*, European Commission Directorate-General for Health and Consumers, Luxembourg, http://ec.europa.eu/health/healthy_environments/docs/env_iaiaq.pdf (accessed 21 June 2014).

Kearns, A. and M. Petticrew (2008), *SHARP Survey Findings: Physical Health and Health Behaviour*, report commissioned by Communities Scotland with the Chief Scientit Office, Crown copyright, Social Research Development Department, Scottish Executive, Edinburgh, www.scotland.gov.uk/Resource/Doc/246083/0069429.pdf (accessed 28 June 2014).

Korsholm, U.S. et al. (2012), "Influence of building insulation on outdoor concentrations of regional air-pollutants", *Atmospheric Environment*, Vol. 54, Elsevier Ltd., Amsterdam, pp. 393-399, (accessed 21 June 2014).

© OECD/IEA, 2014.

Kuholski K., E. Tohn and R. Morley (2010), "Healthy energy-efficient housing: Using a one-touch approach to maximize public health, energy, and housing programs and policies", *Journal of Public Health Management and Practice*, Vol. 16, No. 5 (Suppl.), Lippincott Williams & Wilkins, Hangerstown, pp. S68-S74, http://journals.lww.com/jphmp/pages/articleviewer.aspx?year=2010&issue=09001&article=00011&type=abstract (accessed 18 August 2013).

Laing P. and A. Baker (2006), *The Healthy Housing Program Evaluation: Synthesis and Discussion of Findings*, report prepared for Housing New Zealand Corporation, Housing New Zealand Corporation, Wellington, www.countiesmanukau.health.nz/funded-Services/Intersectoral/docs/HHP-finalreport-year2.pdf (accessed 28 June 2014).Liddell, C. (2014), "Cold homes, pregnant women, and children", *Expert testimony*, NICE (National Institute for Health and Care Excellence), London.

Liddell, C. (2013a), "Tackling fuel poverty: Mental health impacts and why these exist", presentation at the IEA-EEA Roundtable on the Health and Well-being Impacts of Energy Efficiency Improvements, Copenhagen, 18-19 April 2013, www.iea.org/workshop/roundtableonthehealthwell-beingimpactsofenergyefficiencyimprovements.html (accessed 22 June 2014).

Liddell, C. (2013b), *Strategies for Tackling Fuel Poverty among Older People in Northern Ireland*, Age NI (Age Northern Ireland), Belfast, www.ofmdfmni.gov.uk/de/tackling-fuel-poverty-in-ni-liddell-lagdon.pdf (accessed 28 June 2014).

Liddell, C. and C. Guiney (2014), "Improving domestic energy efficiency: Frameworks for understanding impacts on mental health", *Preventive Medicine* (in press), University of Ulster, Coleraine, http://eprints.ulster.ac.uk/29499 (accessed 21 June 2014).

Liddell, C. and S.J.P. McKenzie (2013), *Targeting Those Most in Need: An Areas-based Approach to Tackling Fuel Poverty*, report commissioned by independent research commissioned by OFMDFM (Office of the First Minister and Deputy First Minister), University of Ulster, Coleraine, http://eprints.ulster.ac.uk/27679/1/AWP1_REPORT_FINAL_TYPESET_COPY.pdf (accessed 21 June 2014).

Liddell C., C. Morris and S. Langdon (2011), *Kirklees Warm Zone. The Project and its Impacts on Well-being*, report commissioned by The Department For Social Development Northern Ireland, University of Ulster, Coleraine, www.kirklees.gov.uk/community/environment/energyconservation/warmzone/ulsterreport.pdf (accessed 28 June 2014).

Liddell, C. and C. Morris (2010), "Fuel poverty and human health: A review of recent evidence", *Energy Policy*, Vol. 38, No. 6, Elsevier Ltd., Amsterdam, pp. 2987-2997, www.sciencedirect.com/science/article/pii/S0301421510000625 (accessed 28 June 2014).

Lim S.S. et al. (2012), "A comparative risk assessment of burden of disease and injury attributable to 67 risk factors and risk factor clusters in 21 regions, 1990-2010: a systematic analysis for the Global Burden of Disease Study 2010", *The Lancet*, Vol. 380, No. 9859, Elsevier Ltd., Amsterdam, pp. 2224-2260, www.thelancet.com/journals/lancet/article/PIIS0140-6736(12)61766-8/abstract (accessed 28 June 2014).

LSHTM (London School of Hygiene and Tropical Medicine) and UCL (University College London) (2013), "Monetising the health impact of household energy efficiency improvements", summary documentation for the HIDEEM v2 (Health Impact of Domestic Energy Efficiency Measures version 2) model, May 2013, UCL, London.

Macmillan, A. et al. (2014), "The societal costs and benefits of commuter bicycling: simulating the effects of specific policies using system dynamics modeling", *Environmental Health Perspectives*, Vol. 122, No. 4, National Institute of Environmental Health Sciences, Research Triangle Park, pp. 335-344, http://ehp.niehs.nih.gov/1307250/ (accessed 28 June 2014).

Maidment, C.D.. et al. (2014), "The impact of household energy efficiency measures on health: A meta-analysis", *Energy Policy*, Vol. 65, Elsevier Ltd., Amsterdam, pp. 583-593, www.sciencedirect.com/science/article/pii/S030142151301077X (accessed 28 June 2014).

Marmot Review Team (2011), *The Health Impacts of Cold Homes and Fuel Poverty*, Friends of the Earth and the Marmot Review Team, London.

Matthews, C., et al.(2006), "Influence of exercise, walking, cycling, and overall nonexercise physical activity on mortality in Chinese women", *American Journal of Epidemiology*, Vol. 165, No. 12, Johns Hopkins Bloomberg School of Public Health, Baltimore, pp. 1343-1350, http://aje.oxfordjournals.org/content/165/12/1343.full.pdf (accessed 28 June 2014).

© OECD/IEA, 2014.

McClean Salls, A. et al. (2013), *Rapid HIA: Weatherization Plus Health in Connecticut*, New Opportunities Incorporated, Waterbury, www.healthimpactproject.org/resources/document/WeatherizationPlusHealthConnecticut_Full_Report-1.pdf (accessed 21 June 2014).

Milne, G. and B. Boardman (2000), "Making cold homes warmer: the effect of energy efficiency improvements in low-income homes. A report to the Energy Action Grants Agency Charitable Trust", *Energy Policy*, Vol. 28, No. 6-7, Elsevier Ltd., Amsterdam, pp. 411-424, www.sciencedirect.com/science/article/pii/S0301421500000197 (accessed 28 June 2014).

Milner, J. et al. (2014), "Home energy efficiency and radon related risk of lung cancer: modelling study", *British Medical Journal*, Vol. 348, No. f7493, British Medical Association, London, www.bmj.com/content/348/bmj.f7493.pdf%2Bhtml (accessed 28 June 2014).

Mudarri, D. and W.J. Fisk (2007), "Public health and economic impact of dampness and mold", *Indoor Air*, Vol. 17, No. 3, Danish Technical Press, Copenhagen, pp. 226-235, http://onlinelibrary.wiley.com/doi/10.1111/j.1600-0668.2007.00474.x/abstract (accessed 28 June 2014).

Murray, C. et al. (2000), "Development of WHO guidelines on generalized cost-effectiveness analysis", *Health Economics*, Vol. 9, John Wiley & Sons, Ltd., Hoboken, pp. 235-251, www.who.int/choice/publications/p_2000_guidelines_generalisedcea.pdf (accessed 28 June 2014).

Naughton M.P. et al. (2002), "Heat-related mortality during a 1999 heat wave in Chicago", *American Journal of Preventative Medicine*, Vol. 22, No. 4, Elsevier Science Ltd., Amsterdam, pp. 221-227, www.ajpmonline.org/article/S0749-3797(02)00421-X/abstract (accessed 28 June 2014).

NZ MED (New Zealand government, Ministry of Economic Development) (2011), *New Zealand Energy Strategy: Developing Our Energy Potential and the New Zealand Energy Efficiency and Conservation Strategy*, NZ MED, Wellington, www.med.govt.nz/sectors-industries/energy/pdf-docs-library/energy-strategies/nz-energy-strategy-lr.pdf (accessed 28 June 2014).

Platt, S. et al. (2007), *The Scottish Executive Central Heating Programme: Assessing Impacts on Health*, Social Research Development Department, Scottish Executive, Edinburgh, www.scotland.gov.uk/Resource/Doc/166025/0045176.pdf (accessed 28 June 2014).

Preval N. et al. (2010), "Evaluating energy, health and carbon co-benefits from improved domestic space heating: A randomised community trial", *Energy Policy* Vol. 38, No. 8, Elsevier Ltd., Amsterdam, pp. 3965-3972, www.sciencedirect.com/science/article/pii/S0301421510001837 (accessed 28 June 2014).

Ormandy, D. (ed.) (2009), *Housing and Health in Europe: The WHO LARES Project*, Routledge, London, www.routledge.com/books/details/9780415477352/ (accessed 28 June 2014).

O'Sullivan, K., P. Howden-Chapman and G. Fougere (2011), "Making the connection: the relationship between fuel poverty, electricity disconnection and prepayment metering", *Energy Policy*, Vol. 39, No. 2, Elsevier Ltd., Amsterdam, pp. 733-741, www.sciencedirect.com/science/article/pii/S0301421510007974 (accessed 28 June 2014).

Rabl A. and A. Nazelle (2011), "Benefits of shift from car to active transport", *Transport Policy*, Vol. 19, No. 1, Elsevier Ltd., Amsterdam, pp. 121-131, www.sciencedirect.com/science/article/pii/S0967070X11001119 (accessed 28 June 2014).

RCP (Royal College of Psychiatrists) (2011), *No Health Without Public Mental Health: The Case for Action*, Position Statement PS4/2010, RCP, London, www.rcpsych.ac.uk/pdf/Position%20Statement%204%20website.pdf (accessed 28 June 2014).

Scheer, J. (2013), "Considering health benefits in Irish energy efficiency policy formulation", presentation at the IEA-EEA Roundtable on the Health & Well-being Impacts of Energy Efficiency Improvements, Copenhagen, 18-19 April 2013, www.iea.org/workshop/roundtableonthehealthwell-beingimpactsofenergyefficiencyimprovements.html (accessed 22 June 2014).

Scheer, J. (2012), *Alleviating Energy poverty in Ireland – An Efficient Approach for Future Government Expenditure*, unpublished, available from the author: jim.scheer@seai.ie.

Schweitzer, M. and B. Tonn (2003), "Non-energy benefits of the US Weatherization Assistance Program: a summary of their scope and magnitude", *Applied Energy*, Vol. 76, No. 4, Elsevier Ltd., Amsterdam, pp. 321-335, www.sciencedirect.com/science/article/pii/S0306261903000035 (accessed 23 June 2013).

© OECD/IEA, 2014.

Shrubsole, C. et al. (2014), "100 unintended consequences of policies to improve the energy efficiency of the housing stock", *Indoor and Built Environment*, 12 March 2014, Sage Publications (online), http://ibe.sagepub.com/content/early/2014/03/12/1420326X14524586.full.pdf+html (accessed 28 June 2014).

Slotsholm (2012), *Socio-economic Consequences of Better Air Quality in Primary Schools*, report prepared by Slotsholm A/S in collaboration with Velux A/S and the Technical University of Denmark, Slotsholm A/S, Copenhagen, https://www.velux.com/ar-DZ/Daylight/ventilation/facts_ventilation/did_you_know/Documents/socio-economic-consequences-og-better-air-quality-in-primary-schools_slotsholm_uk.pdf (accessed 28 June 2014).

Staddon, P., H. Montgomery, M. Depledge (2014), "Climate warming will not decrease winter mortality", *Nature Climate Change*, Macmillan Publishers Limited, Vol. 4, No. 3, pp. 190-194.

Stearn, J. (2012), "Empowering consumers in vulnerable positions – civil society and the market place", Consumer Futures website, www.consumerfutures.org.uk/blog/consumer-focus-empowering-consumers-in-vulnerable-positions-civil-society-and-the-market-place (accessed 28 June 2014).

Sunikka-Blank, M. and R. Galvin (2012), "Introducing the prebound effect: the gap between performance and actual energy consumption", *Building Research & Information*, Vol. 40, No. 3, Routledge, London, pp. 260-273, www.arct.cam.ac.uk/Downloads/introducing-the-prebound-effect-the-gap-between-performance-and-actual-energy-consumption-minna-sunikka-blank-and-ray-galvin (accessed 28 June 2014).

Telfar-Barnard, L. et al. (2011), *The Impact of Retrofitted Insulation and New Heaters on Health Services Utilisation and Costs, Pharmaceutical Costs and Mortality: Evaluation of Warm Up New Zealand: Heat Smart*, report to the MED (Ministry of Economic Development), MED, Wellington, www.motu.org.nz/ (accessed 21 June 2014).

Tetra Tech, Inc. and Massachusetts Program Administrators (2011), *Massachusetts Special and Cross-sector Studies Area, Residential and Low-income Non-energy Impacts (NEI) Evaluation*, final report prepared for Massachusetts Program Administrators, Tetra Tech, Inc., Madison, www.rieermc.ri.gov/documents/evaluationstudies/2011/Tetra_Tech_and_NMR_2011_MA_Res_and_LI_NEI_Evaluation(76).pdf (accessed 28 June 2014).

Thomson, H. et al. (2013), "Housing improvements for health and associated socio-economic outcomes", Cochrane database of systematic reviews, John Wiley & Sons Ltd., Hoboken, http://onlinelibrary.wiley.com/doi/10.1002/14651858.CD008657.pub2/pdf (accessed 28 June 2014).

Tod A.M. et al. (2012), "Understanding factors influencing vulnerable older people keeping warm and well in winter: a qualitative study using social marketing techniques", BMJ Open 2012;2:e000922. doi:10.1136/bmjopen-2012-000922, http://bmjopen.bmj.com/content/2/4/e000922.full.pdf+html (accessed 28 June 2014).

Tonn, B., et al. (2014), *Health and Household-Related Benefits Attributable to the Weatherization Assistance Program*, confidential draft, report prepared for the US DOE (United States Department of Energy), Oak Ridge National Laboratory, Oak Ridge, http://weatherization.ornl.gov/pdfs/ORNL_CON-484.pdf (accessed 28 June 2014).

UK DCLG (Department for Communities and Local Government, United Kingdom Government) (2006), "Reducing the risks: the housing health and safety rating system", Ref: 05HD03402/H, Office of the Deputy Prime Minister, London, www.gov.uk/government/publications/reducing-the-risks-the-housing-health-and-safety-rating-system (accessed 21 June 2014).

US EPA (United States Environmental Protection Agency) (2011), *Healthy Indoor Environment Protocols for Home Energy Upgrades: Guidance for Achieving Safe and Healthy Indoor Environments During Home Energy Retrofits*, US EPA, www.epa.gov/iaq/pdfs/epa_retrofit_protocols.pdf (accessed 28 June 2014).

US NCHH (US National Center for Healthy Housing) and American Public Health Association (2013), *National Healthy Housing Standard*, US NCHH, Columbia, www.nchh.org/Portals/0/Contents/NHHS_Full_Doc.pdf (accessed 6 May 2014).

Vandentorren, S. et al. (2012), "The impact of heat islands on mortality in Paris during the August 2003 heat wave", *Environmental Health Perspectives*, Vol. 120, No. 2, National Institute of Environmental Health Sciences, Research Triangle Park, pp. 254-259, www.ncbi.nlm.nih.gov/pmc/articles/PMC3279432/pdf/ehp.1103532.pdf (accessed 28 June 2014).

© OECD/IEA, 2014.

Walker, R. et al. (2013), "Area-based targeting of fuel poverty in Northern Ireland: An evidenced-based approach", *Applied Geography*, Vol. 34, Elsevier, B.V., Amsterdam, pp. 639-649, www.sciencedirect.com/science/article/pii/S0143622812000288 (accessed 28 June 2014).

World Bank and SEPA (State Environmental Protection Administration), P. R. China (2007), *Cost of Pollution in China: Economic Estimates of Physical Damages*, World Bank, Washington DC, http://siteresources.worldbank.org/INTEAPREGTOPENVIRONMENT/Resources/China_Cost_of_Pollution.pdf (accessed 28 June 2014).

Wilkinson, R.G. (2005), *The Impact of Inequality: How to Make Sick Societies Healthier*, Routledge, London.

WHO (World Health Organization) (2014), *Ambient (Outdoor) Air Quality and Health,* Fact sheet N°313 (Updated March 2014), www.who.int/mediacentre/factsheets/fs313/en/ (accessed 15 June 2014).

WHO (2011a), *Health in Green Economy. Health Co-benefits of Climate Change Mitigation – Housing Sector,* WHO, Geneva, www.who.int/hia/hgehousing.pdf (accessed 28 June 2014).

WHO (2011b), *Burden of Disease from Environmental Noise: Quantification of Healthy Life Years Lost in Europe,* Regional Office for Europe, WHO, Geneva, www.euro.who.int/__data/assets/pdf_file/0008/136466/e94888.pdf (accessed 28 June 2014).

WHO (2009), *Global Health Risks: Mortality and Burden of Disease Attributable to Selected Major Risks*, WHO, Geneva, www.who.int/healthinfo/global_burden_disease/GlobalHealthRisks_report_full.pdf (accessed 28 June 2014).

WHO (2006), *Housing, Energy and Thermal Comfort: A Review of 10 Countries within the WHO European Region*, Regional Office for Europe, WHO, Geneva, www.euro.who.int/__data/assets/pdf_file/0008/97091/E89887.pdf (accessed May 2014).

Woodcock, J. et al. (2009), "Public health benefits of strategies to reduce greenhouse-gas emissions: urban land transport", *The Lancet,* Vol. 374, No. 9705, Elsevier Ltd., Amsterdam, pp. 1930-1943, www.thelancet.com/journals/lancet/article/PIIS0140-6736(09)61714-1/abstract (accessed 28 June 2014).

© OECD/IEA, 2014.

Chapter 5

Industrial sector impacts of energy efficiency

Key points

- *Energy efficiency investments can be strategic for industry, supporting core business activities by helping to reduce costs, increase value and mitigate risk. Quantification and monetisation of the multiple benefits can enable benefit-cost analysis, thereby strengthening the business case for energy efficiency.*
- *Improved energy efficiency can deliver multiple benefits across the industry value chain, leading to enhanced competitiveness, more cost-efficient production, and reduced operation and maintenance (O&M) costs. It can also lower the costs of environmental compliance and improve the working environment for employees.*
- *Integrating these broader benefits into energy efficiency assessments can deliver surprising results for the criteria executives use when making investment decisions. In multiple studies, a more comprehensive assessment of benefits reduced the payback period by half (e.g. from four years to less than two).*
- *Even small businesses have much to gain: in measuring the impacts from a series of production component replacements, a secondary lead producer in Peru calculated energy cost savings of USD 1 850 per year. The associated increase in production delivered a value of USD 16 980 per year – almost ten times higher.*
- *A key challenge in this area is better aligning energy efficiency policy initiatives with industry needs and goals – including the need to consider multiple stakeholder perspectives. This implies stronger collaboration to address issues such as data sharing and confidentiality.*

Introduction: Emerging evidence of industrial sector impacts

Industry[1] accounts for one-third of global final energy demand. International Energy Agency (IEA) analysis shows that while industrial energy efficiency is improving, large potential remains untapped. A key challenge is that industry is highly heterogeneous. There are thousands of industrial processes and countless ways in which energy efficiency projects can be designed and implemented – ranging from replacing one piece of equipment to complete facility retrofit and modernisation.

The impacts of industrial energy efficiency measures are routinely calculated only in terms of energy demand reduction, and sometimes in terms of greenhouse gas abatement. This IEA review of existing studies shows that energy efficiency measures in industry can provide a range of additional direct benefits for businesses i.e. multiple benefits.[2] These

1 In this context, the IEA uses the term "industry" to include industrial and commercial companies ranging from small- and medium-sized enterprises to large corporations.

2 In other literature, these impacts have been variously labelled "co-benefits", "ancillary benefits" and "non-energy benefits" (NEBs) – and are often used interchangeably with "multiple benefits". The IEA uses the term multiple benefits, which is broad enough to reflect the heterogeneous nature of outcomes and to avoid pre-emptive prioritisation of various benefits; different benefits will be of interest to different stakeholders.

© OECD/IEA, 2014.

include: enhanced productivity and competitiveness; reduced costs for environmental compliance, O&M and waste disposal; extended equipment lifetime; improved process and product quality; and improved work conditions and decreased liability. The chapter explores the topic from both industry and policy-maker perspectives, with a view to identifying ways for policy to stimulate industry engagement.

A new body of evidence is emerging to show that investment in industrial energy efficiency projects can generate additional positive impacts for individual companies and for the economy (Pye and McKane, 2000). These benefits, however, are not usually properly quantified and valued. Including these additional benefits in decision-making processes can improve the attractiveness of energy efficiency investments and measures in industry. From a policy-making perspective, two main considerations support the case for broadening assessment of energy efficiency programmes and policies to include these wider impacts:

- Better understanding of the wider benefits may strengthen the business case for energy efficiency in companies, which in turn could stimulate the implementation of energy efficiency measures and the achievement of programme objectives.
- Increased uptake by industry could generate more public funding for industrial energy efficiency policies, particularly if it can be shown that industrial energy efficiency policies contribute to wider policy objectives such as environmental protection, innovation and economic development.

Evidence that has emerged to date reveals the scope of potential impacts and glimpses of their potential value for reducing costs, increasing value and mitigating risk in industry. But consensus is lacking on methodologies that can be used to systematically identify, quantify, monetise and report on the wider multiple benefits. Gathering the data and developing methodologies to improve understanding of the value (from monetary or other perspectives) of these multiple benefits could, in the long run, bring energy efficiency into the realm of strategic importance for companies.

In addition to providing an overview of the full range of benefits that can arise from energy efficiency policies and measures targeting the industrial sector, this chapter offers guidance on their quantification. It outlines the steps that can be used for initiating work in this area and identifies opportunities for research and initiatives that would expand the knowledge base and prompt further development of methods.

Strengthening the business case for industrial energy efficiency

An important consideration is that while energy efficiency policy makers may see a certain logic in distinguishing between general projects and energy efficiency projects, this is not necessarily the case within industry. Improved energy efficiency tends to be viewed by industry as a by-product of initiatives undertaken for business reasons; it is rarely a targeted aim (Box 5.1). The exception is when energy efficiency projects are linked to specific government targets or energy efficiency reporting requirements, in which case they may be viewed by industry as a burden on time, resources and finances.

It is often claimed that energy efficiency is not part of core business, and thus not of strategic importance to companies. Considering a wider range of benefits from energy efficiency, and demonstrating their direct impact on business processes and productivity, can strengthen the business case for energy efficiency investments. Evidencing the role energy efficiency improvements can play in other types of modernisation projects might help raise industry interest in exploring available technologies, methodologies or measures.

© OECD/IEA, 2014.

Box 5.1 The primary objective: Industrial optimisation or upgrading projects?

Worsley Alumina, an Australian aluminium producer, initiated a system optimisation project with the stated aim of reducing energy demand. Ultimately, the measure delivered additional benefits in the form of reduced need for operator intervention (reduction of workload and operator error), improved system stability and reliability, lower maintenance needs and fewer charge-outs. These multiple benefits enabled the company to increase production by 3 000 tonnes of aluminium per year, having a commercial value of USD 6 million per year (USD/yr) (given a sales price of USD 2 000 per tonne) (DRET, 2013).

Understanding what drives business investment decisions is fundamental to efforts to stimulate additional investments in energy efficiency in an industrial context. Return on investment is, in most cases, a primary consideration; in some cases, however, the strategic nature of an investment may have a greater influence. Energy efficiency can be strategic: it supports core business activities by helping to reduce costs, increase value (including added value through innovation and diversification) and mitigate risk (Box 5.2).

Box 5.2 Key drivers for business-led energy efficiency

Companies usually prioritise investments that contribute to improving their bottom line – i.e. delivering a return on investment by improving profit margins. Energy efficiency investments can contribute to the strategic priorities of businesses in three key ways:

- **Reduce costs**: in addition to reducing the amount of money spent on energy, energy efficiency can lead to lower expenses for water, materials, equipment repair and maintenance, etc.
- **Increase value** (including added value through innovation and diversification): energy efficiency can help companies generate value by improving product quality; in some cases, it can help companies win supplier contracts and access new markets. Companies can also gain market share by highlighting their commitment to energy efficiency in advertising and branding.
- **Mitigate risk**: energy price volatility is one of several risks that energy efficiency can help mitigate; as energy and fuel prices influence the cost of products, lower energy consumption can help to keep down costs and create competitive advantage. Energy efficiency projects can also contribute to improving processes and reducing the risk of downtime. Some energy efficiency projects can improve worker safety, thus reducing risks of negative health impacts or accidents. At present, energy efficiency implementation is sometimes associated with *increased* risk – for example, risks stemming from the need to change processes or impacts of equipment replacement on processes. More well-rounded assessments will contribute to better management of such risk and help dispel unfounded risk perceptions.

Source: Cooremans, C. (2012), "Investment in energy efficiency: Do the characteristics of investments matter?", *Energy Efficiency*, Vol. 5, No. 4, Springer Netherlands, Houten, pp. 497-518.

© OECD/IEA, 2014.

Energy managers in industry engaged in energy efficiency typically focus on energy cost reduction when putting forth the business case for investment, often neglecting the importance of value creation and risk mitigation. The other potential benefits are not yet part of their key performance indicators (KPIs), in part because they may lack access to the data needed to quantify other benefits. They may also lack the know-how and capacity to identify and quantify these benefits. This has two implications:

- The full costs and benefits of energy efficiency are not assessed or measured, or are not used to inform decision making (the focus on energy cost savings falls short of capturing the full range and value of costs savings).
- The project is presented or communicated to decision makers in a way that fails to highlight its strategic nature (aspects of value creation and risk mitigation are neglected).

Incentivising industry action on energy efficiency

Energy efficiency investments are frequently perceived as minimally to moderately strategic. While generally accepted as contributing to the cost constituent, this perspective may not be particularly powerful in motivating companies towards investing in energy efficiency projects and/or energy management systems (EnMSs).

It is also the case that other costs – for labour, capital and other resources – tend to eclipse energy costs in all but a limited range of energy-intensive process industries. So potential energy demand reduction is not viewed as a high priority. In fact, research indicates companies typically demand shorter payback periods for energy efficiency investments than for other investments. Yet in many cases, the payback periods for energy efficiency is two years or less, equivalent to a discount rate of considerably more than 50% (McKinsey Global Institute, 2007).

For many companies, cost savings will be perceived as marginal compared to strategic value creation. Priorities differ among companies, and even priorities within a company will shift according to business objectives at a given point in time. Ultimately, investment decisions require a high level of alignment with strategic aims: reducing costs may be an ongoing concern and a high priority, but is often considered as operational more than strategic.

Strengthening the capacity to identify and quantify the wider benefits of energy efficiency projects in terms of cost reduction, value generation and risk mitigation to support their inclusion in investment assessments could significantly raise the profile of energy efficiency.

Taking account of stakeholder perspectives

The bottom line is often "all-important" in industry, not only because of the drive for profits but because companies have numerous stakeholders to consider when making investment decisions. In addition to the energy manager pitching energy efficiency measures to executives, the executives then need to think about how to market the additional benefits of energy efficiency to diverse stakeholders and make it relevant to their particular set of concerns.

Incorporating a fuller range of impacts into assessment frameworks can be useful to a number of industry stakeholders (Table 5.1). For company management, the value of enhanced impact information will be in supporting decision making and improving benefit-cost assessments that enable deeper energy demand reduction and maximise other benefits. Shareholders will likely respond positively to information on measures that will lead to improved return on investments. What particular stakeholders deem valuable will vary among different types of companies, sectors and countries.

© OECD/IEA, 2014.

Table 5.1 Getting stakeholder buy-in to the business case for energy efficiency

Type of stakeholder	Potential benefits of assessing wider impacts of industrial energy efficiency
Company-level management	More comprehensive assessment of benefits and costs of energy efficiency investments or projects. Improved benefit-cost ratios, with shorter payback periods for energy efficiency projects. Enable deeper energy demand reduction. Improved risk profile. Enhanced corporate sustainability reporting (CSR). Opportunity to use lessons learned and experience gained from this area to incorporate a multi-benefits assessment methodology for every type of project. Improved industrial competitiveness.
Personnel engaged in energy management within companies	Ability to make a better business case for investments in energy efficient technologies or other types of projects that promote energy efficiency. Enhanced ability to gain management interest in energy efficiency projects. Better understanding of the interconnectedness between energy and other company resources. Improved ability to report on energy performance (as required within some EnMSs, notably ISO 50001).
Shareholders	Better return on investment. Increased understanding of how key resources affect and shape business. Richer understanding of business investment performance. Enhanced CSR.
Financial sector	Improved bankability of energy efficiency projects (and any projects with an energy efficiency dimension). Risk reduction.
Programme or policy level	Enhanced ability to provide more comprehensive assessment of results and impacts from programmes or policies (beyond traditional cost-effectiveness assessments). Provide better/more comprehensive public accountability on policy outcomes. Enhanced ability to design energy efficiency policy instruments that meet other policy objectives (beyond energy demand reduction). Generate new knowledge that can be used to reshape public policy. Help justify funding and additional resources, programme continuation or expansion by providing assessments of broader impacts that meet policy objectives beyond energy efficiency. Focus on wider impacts can enhance other programme objectives. Improved ability to engage industry (e.g. industrial supply chains) in programmes by showing quantified benefits beyond cost reductions from reduced energy use. Help communicate the strategic value of EnMSs when integrated with other business tools to help improve the robustness of economic assessments for industrial projects.
Energy planning and strategic level	Justification for investing in policies to promote energy efficiency. Improved decision-making basis for allocating resources – e.g. energy efficiency, new generation or other measures. Deeper understanding of economic growth drivers.

Source: Unless otherwise noted, all material in figures and tables in this chapter derives from IEA data and analysis.

Key point *Energy efficiency projects can meet the diverse interests of a wide range of stakeholders.*

© OECD/IEA, 2014.

The range of industrial sector impacts

Identifying the multiple benefits that may be linked to energy efficiency measures in industry could enhance the business case for action. It could serve as a basis for assessing which impacts are of relevance in particular contexts and to start tracking their occurrence in practice. Hundreds of different benefits to industry have already been identified in past studies and surveys of energy efficiency project implementation, making it challenging to produce a definitive list of the most important ones. What has become clear is that the relative importance of different types of benefits will vary depending on the sector, type of company and company priorities. One way to deal with the complexity is to organise the benefits into a manageable number of general categories based on which areas they impact – competitiveness, production, O&M, working environment or environment (Table 5.2).

The categories proposed below are not definitive or discrete. Some benefits can fit into several categories, and causality and inter-linkages exist among different types of benefits. For example, an energy efficiency project that delivers health and safety benefits will ultimately also contribute to reducing corporate risk. Production improvements can reduce waste, which can be categorised as an environmental benefit.

Many activities that improve energy efficiency also provide better control over processes. Energy efficient practices can ensure that thermal resources are applied at the right temperature, for the right duration and in correct proportion to raw materials. In addition to reducing the energy consumed per unit of production, this control reduces a facility's scrap rates. Control also provides reliability, which means less downtime and less downtime can improve productivity —thus generating more revenue (NAM, 2005).

Establishing a clear causal link between an energy efficiency measure and a specific set of outcomes may be challenging since, in many cases, numerous variables (not just the energy efficiency measure) may influence a specific outcome. Initially, it may be wise to focus on benefits for which it is relatively easy to establish a link. While causality tests may be helpful, using common sense may be sufficient to make reasonable assumptions.

Table 5.2 Company-level benefits from industrial energy efficiency projects

Benefit	Description
Competitiveness	
Ability to enter new markets/increased market share	Overcoming technical barriers to trade or overcoming market perceptions or resistance (e.g. perception about carbon dioxide [CO_2] footprints).
	Expanded capacity or new product features that enable entrance in new markets.
Reduced production costs	Reduced costs per unit or enabling the company to access and capitalise on a new complementary or substitute factor of production and in doing so opening up new opportunities for growth.
Deferred plant capital investments	Optimising processes or upgrading equipment or extended equipment lifetime can defer the need for capital costs in replacing equipment. Optimising processes for energy efficiency can also lead to situations where certain equipment is redundant.
Corporate risk reduction	Mitigation of corporate risk through reducing liabilities and helping to achieve or go beyond current regulatory requirements.
Improved reputation, corporate image	Improved corporate image through publicising energy efficient (sustainable) business. Improvement of corporate image through CSR that incorporates the wider range of benefits (both private benefits and public benefits).
	Better brand reputation through product or service quality improvements.

© OECD/IEA, 2014.

Benefit	Description
Production	
Capacity utilisation	More efficient equipment or processes can lead to shorter process times and use of lower cost factors of production (labour and materials), which can lower production costs and enable higher product output.
Improved product quality	Downstream improvements in reductions in product defects and warranty claims as well as contributing to enhanced brand reputation.
Increased product value	Improved quality and consistency contributes to added value which in turn can contribute to enhanced brand reputation.
O&M	
Improved operation	Improved operation and process reliability leads to reduced equipment downtime, reduced number of shutdowns or system failures and can entail reduced process time (which can contribute to increased productivity), process optimisation can also reduce staff time required to monitor and operate a processing plant is therefore reduced, which reduces overhead costs.
Reduced need for maintenance	Energy efficiency projects can lead to investments in new equipment, system optimisation, optimisation or change of processes which in turn can lead to lower maintenance requirements (or avoidance of extraordinary maintenance), reduced costs for maintenance, reduced cost for maintenance materials.
Working environment	
Improved site environmental quality	Improved work environment from improved thermal comfort, lighting, acoustics and ventilation. Improved conditions can help retain and attract skilled staff. Improved work conditions and work environment can increase labour output.
Increased worker health and safety	Process improvements and equipment upgrades implemented as part of energy efficiency projects can reduce the risk and incidence of work-related accidents or negative impacts on worker health. Such improvements can lead to reduced health insurance costs and medical expenses (as well as reduce corporate risk – liability in case of accidents).
Environment	
Reduction of air pollution and emissions	Reducing energy use or optimising processes can reduce sulphur oxides (SO_x), nitrogen oxides (NO_x), carbon monoixide (CO), chlorofluorocarbons (CFCs), hydrofluorocarbons (HFCs), as well as CO_2 emissions and associated credit or reduced compliance costs. Process changes reduce combustion and process emissions can be important to industry when there are regulatory or compliance issues and associated cost savings include avoiding fines or taxes.
Solid waste reduction	Reducing waste streams through e.g. production improvements, product redesign, improved operation result in less waste, which reduces waste disposal/abatement costs and input materials purchase cost.
Waste water reduction	Process optimisation, improved operation, improved maintenance can reduce water needed to run processes or water needed for cleaning purposes. Reducing wastewater has environmental benefits but can also entail reduced costs for wastewater treatment.
Reduction of input materials, e.g. water	Reduction of input materials reduces upstream environmental impacts from extraction, processing and transport.

Notes: Categories and benefits are not listed in order of importance. This is not an exhaustive list.

Key point *The range of energy efficiency benefits in industry can be organised into five main categories.*

© OECD/IEA, 2014.

Box 5.3 Energy efficiency can reduce equipment damage

A Danish company initiated an energy efficiency project to reduce energy demand in the process of producing liquid gases. Using a combination of an ozone unit and a sand filter, it was possible to reduce the temperature of cooling water.

Implementation showed energy savings of 153 000 kilowatt hours per year (kWh/yr) corresponding to annual savings of USD 12 000. This process improvement also reduced the amount of required process chemicals (giving savings of USD 50 000/yr), reduced the need for corrosion inhibitors (saving USD 12 000/yr) and reduced corrosion damage (valued at USD 20 000/yr).

In addition, the company noted reduced labour costs, less down time, reduced negative environmental impacts and an improved working environment (Gudbjerg, Dyhr-Mikkelsen and Monrad Andersen, 2014).

A vital consideration is that benefits vary in terms of the time horizon on which they occur. Some are immediate, such as the reduction of cost for input materials; others, such as expanded market share due to enhanced product quality, may occur at some point in the future once the market reacts to the changes in product quality. Determining appropriate timeframes for measuring impacts and taking into consideration any changes in the value of savings (or avoided costs) over time will be important. For instance, it can be expected that cost savings from reduced maintenance costs will decrease as the equipment or system ages. Similarly, if the value of product outputs increases, then the value of production-related benefits could increase over time. In some cases, some costs could increase initially (e.g. maintenance costs upon installing new equipment), while in a longer-term perspective the same costs could decline. Thus, in certain circumstances, a longer timeframe may be needed to capture the long-term benefits of an energy efficiency measure.

Benefits that are likely to be of highest interest to industry players are those which:

- drive a rapid return on investment
- contribute to cost reduction, value generation and risk mitigation
- are relevant to industry and expected to have a relatively high monetary value
- can be clearly linked to implementation of specific projects
- for which good information and data are available and accessible
- have the potential to increase access to energy efficiency finance.

Work to date indicates the value of these additional benefits can be in the range of 40% to 50% of the value of the actual energy demand reduction per measure (or as much as 2.5 times depending on the size and the context of the investment) (Lilly and Pearson, 1999; Pearson and Skumatz, 2002).[3]

Several broader socio-economic benefits relate to improved energy efficiency in industry, such as reductions in local and global pollution, employment creation, stimulation of new business sectors (e.g. energy efficiency service and technology providers), improved international competitiveness and enhanced energy security. These impacts are highlighted in Chapters 1 and 2.

3 The value of benefits varies between different types of projects and measures implemented, between companies and between sectors. It is not possible to assume that all types of projects would achieve benefits in addition to energy efficiency in this range. Some energy efficiency projects and measures can also lead to negative impacts, for example, disruptions to production processes or downtime. Quantifying these is also essential for company-level decision making as part of risk management.

© OECD/IEA, 2014.

Box 5.4

Energy efficiency measures boost production in small manufacturing

As part of their efforts to improve energy efficiency, Metalexacto (a small, secondary lead producer based in Peru) replaced a burner, optimised the fuel mix used, changed refractory bricks and installed a hood on the furnace. This enabled an increased extraction of lead – in the range of 34.7 tonnes per year at a value of USD 16 980/yr. Meanwhile, the energy demand reduction achieved had a value of only USD 1 850/yr (UNIDO/UNEP, 2010b).*

Most of the added value of this project comes not from the energy demand reduction but from the additional benefits stemming from the energy efficiency improvements.

* While this example indicates that other benefits can match or even surpass the value of those from energy demand reduction, it should be noted that the value of saved energy is highly dependent on local energy market conditions, which may change over time. Consequently, the value ratio to other benefits is also highly dependent on local energy market conditions.

To properly account for the value of energy efficiency in industry, both positive and negative impacts should be assessed. Negative aspects could include, for example, decreases in productivity due to down time and personnel training for equipment upgrades. Quantifying negative impacts involves similar challenges as quantifying benefits, but it is essential to take these into account to ensure the robustness of any assessment. In many cases, but not always, the positive effect will outweigh the negative one.[4]

Methodological approaches

Because so few studies have been undertaken in this area, methodologies for quantifying wider benefits from energy efficiency measures in industry are still at the inception stage. This section outlines some approaches and seeks to draw out key issues that need to be considered in the process. Work in this area remains too exploratory to start to suggest a standardised methodology, but a basis for pursuing future work does already exist.

Early estimates on the broad order of magnitude for the value of benefits could already contribute to improving the attractiveness of energy efficiency projects. Approaches within accepted financial and economic analysis frameworks can provide a starting point for collecting and processing data on industrial benefits. Initial data collection efforts can, in turn, be used to increase understanding of the scope of various benefits and further refine approaches, working towards an increasingly robust methodological framework.

Several academic studies that have assessed industrial benefits provide a source for early lessons on methodological approaches. Typically, approaches involve retrospective analysis using a range of methods to identify and quantify benefits, which then enable an assessment of the overall impact (Box 5.5).

Collecting data

When applying a multiple benefits approach within companies, decisions about what benefits to focus on and what approach to use to quantify them will largely be determined by three factors: the availability of data, the time required to make calculations and the time required to report outcomes.

4 From a policy maker perspective, it is important to be sensitive to possible negative impacts and to assess them as part of more comprehensive evaluations of energy efficiency programmes.

© OECD/IEA, 2014.

Box 5.5 Early assessments of the multiple benefits of industrial energy efficiency

In an investigation of 70 industrial case studies, Worrell et al. (2003) initially identified and described productivity benefits associated with a given energy efficiency measure. These benefits were quantified to the extent possible; then, using identified assumptions needed to translate the benefits into monetary terms, the monetary value of productivity benefits was calculated. The values were then incorporated into conservation supply curves (CSCs).

Lilly and Pearson (1999) evaluated benefits on the basis of five case studies. To carry out their assessment, they met with energy managers, plant managers, and O&M personnel to determine the best approach (considering the availability of data) and develop an evaluation plan. After data collection and analysis, regression models were developed and used to draw out the various impacts of energy efficiency measures. Additional benefits were found to account for 24% of overall benefits of energy efficiency measures implemented.

Finman and Laitner (2002) analysed 77 case studies to get an indication of the value of the additional benefits attributable to energy efficiency in a manufacturing setting. Of the total number of cases, 52 included a monetised estimate of both energy cost savings and additional benefits. Based on energy cost savings alone, project paybacks in aggregate were 4.2 years. With additional benefits included, the aggregate payback fell by more than half to 1.9 years.

This decrease in payback period from 4.2 to 1.9 years also emerged in other studies when additional benefits were included. The net financial savings from the studied energy efficiency measures varies greatly, ranging from 0.03% to 70% of the total savings upon the inclusion of additional benefits (Worrell et al., 2003). Other studies have evaluated the effect of non-energy savings by calculating payback periods for two scenarios, one incorporating additional benefits and one with energy cost savings only. By including additional benefits into the CSC, the payback period was reduced by 31% from 1.43 years to 0.99 (Lung et al., 2005).

Hall and Roth (2003) conducted a study of 210 companies. To assess the full range of benefits from energy efficiency projects, they carried out in-depth interviews with management and staff involved in the implementation of energy efficiency measures. Through the interview process, Hall and Roth identified indicators and metrics that they were able to rank according to importance. They then integrated the value of these benefits into benefit-cost ratios.

Pearson and Skumatz (2002) assessed the impact of a commercial industrial energy and water efficiency programme. Using interviews, they asked participants to list additional benefits and estimate their value in terms of being larger or smaller than the value of the achieved or expected energy cost savings. The relative similarity of energy efficiency improvements identified through the programme made it possible to pinpoint the types of benefits that delivered the greatest value and to assign indicative values. While these figures cannot be directly used to assess the value of benefits achieved via other programmes, the study shows that it is possible to make a reasonable assessment of the order of magnitude of savings, as well as gain insights that can be used to improve energy efficiency programme design.

These studies show that even if there is no single approach to assessing benefits, different approaches can be used depending on the context, the availability of data and, ultimately, the purpose of the assessment.

© OECD/IEA, 2014.

Policy makers strive to acquire as much data as possible to inform policy design, development and implementation. A solid data foundation can help generate a higher level of interest in energy efficiency within industry and in the broader policy community. Realistically, there are limits to how much pressure policy makers can place on companies in terms of reporting data or outcomes. A pragmatic and strategic approach is needed to identify priorities at the outset.

The easiest course of action is to base quantification on actual data for savings or value. For some types of benefits, data and information may be already available within a company; it is important to investigate what data various departments are already collecting through other processes that could be of use. This approach could both expedite the process and save resources, but is not always possible.

From the policy-maker perspective, two main aspects warrant consideration: i) how to facilitate or stimulate data collection; and ii) how to incentivise companies to share or report data. The data needed to enable policy makers to assess the impacts of industrial energy efficiency are not routinely reported to governments and may not even be collected at the company level. Systems for data collection and reporting before, during and after energy efficiency interventions should be established. These should seek to avoid placing an unnecessary burden on companies and enhance the comparability of data derived from different companies, by placing priority on ease of reporting (including clear guidelines) and building synergies with existing reporting systems. An ongoing process of developing methods for extrapolating new and improved data on multiple benefits might be supported by national energy efficiency agencies, or other entities that administer industrial energy efficiency programmes. These bodies play an important role in promoting knowledge exchange and data collection.

Some international efforts are currently underway to address the need for data and information in the area of energy efficiency for industry. Although these programmes remain limited, they do provide a useful foundation (Box 5.6).

Box 5.6 Initiatives to promote data collection and sharing

At present, two main initiatives are taking steps to build platforms for collecting and sharing data, information and experiences in the area of industrial energy efficiency benefits, led by the Global Superior Energy Performance (GSEP) and the Energy Management Action Network (EMAK).

The Energy Performance Database, developed through collaboration with GSEP member countries under the auspices of the Clean Energy Ministerial, seeks to develop a strong, data-driven business case for implementing energy management business practices. The database provides an opportunity to make connections between national energy management programmes and policies and the benefits that facilities and companies experience due to implementing an EnMS. Initial outputs will include a set of key findings that can be used to: support the business case for implementing an EnMS, provide guidance for the developing EnMS programmes and policies, and encourage growth of the database by demonstrating its value. As more data become available, outputs from the Energy Performance Database will also provide valuable information on industrial energy use and consumption, the role of operational changes and technology upgrades in improving energy performance, and the impact of energy management practices on additional benefits (such as maintenance and productivity).

EMAK is a network that brings together policy makers and private energy managers to share best practices and identify promising approaches in developing effective industrial energy efficiency policies. The network also ensures the implementation of energy efficiency measures in industry. EMAK organises workshops and webinars and is creating a web-based platform to facilitate information sharing.

© OECD/IEA, 2014.

Once priority benefit areas have been identified, the data requirements for beginning detailed assessment of them will need to be considered, and plans put in place for building up an information base. Most of the data and information that is relevant to assessing the impact of energy efficiency policy in industry is held by the industry players themselves. To acquire the data needed to assess the relative value of different policy options, energy efficiency programmes need to build in mechanisms by which companies report on impacts and outcomes. An important consideration for policy makers is to ensure that data requests align with the needs and timeframes of companies surveyed. In addition, a high degree of confidentiality is needed to protect intellectual property rights and competitive information (Box 5.7).

Box 5.7 Consider confidentiality

A range of costs and benefits related to energy efficiency are being assessed within industry, but information on the outcomes of these assessments is typically not shared outside the company. There may be restrictions in terms of access of data needed to quantify certain benefits in industry, e.g. benefits related to employee health and productivity. This may influence the choice of benefits to be measured. Policy makers will need to address this issue by providing sufficient security in relation to confidentiality and competitiveness issues.
As with other industry reporting, this can be achieved via appropriate confidentiality agreements or through publication of aggregated figures only.

Companies (as opposed to governments) have different reasons for assessing the wider impacts of energy efficiency. *Ex ante* assessments will generally be undertaken to inform investment decisions, but *ex post* assessment may be less rigorously pursued if there is no compelling external incentive (e.g. reporting to government, sustainability reporting, or reporting in connection to energy performance contracts or as part of financial assessments). In fact, it does make sense for companies to track impacts from project implementation so that results can be fed back to inform new investment decisions (Box 5.8).

Box 5.8 A company-level *ex ante* assessment of additional benefits

SSAB, a Swedish company producing high-strength steel, conducted a project in hydraulic system optimisation. The total project cost was USD 53 000 and led to a 58% reduction in energy use, equal to a monetary saving of USD 18 000/yr.

After the project was finalised, SSAB became interested in understanding the wider benefits of the investment. The identified additional benefits were: reduced wear on pumps, motors and the overall installation; lower maintenance costs of filter and oil change; and a reduced need for cooling of the oil. The estimated value of these benefits combined is USD 30 000/yr.

Adding the savings from all benefits brings the total to USD 48 000/yr, meaning a payoff time of less than two years. SSAB also noted an improved reliability in operations.

© OECD/IEA, 2014.

Understanding how to motivate companies to collect such data is a first step towards improved information on the wider impacts of industrial energy efficiency. Equally important is learning how they could be incentivised to share this information with policy makers. Focus should be placed on eliciting information on which benefits are of strongest interest to industry and for which it is possible to gather necessary data and information.

Optimising the data collection demands and processes is critical. It does not make sense to expend large resources on collecting and processing data if the value of the benefit is expected to be marginal. Nevertheless, caution should be applied in making this selection, so as not to exclude potentially important benefits.

The approach to data collection depends on the needs and means of companies, as well as site-specific factors. Policy makers need to take into consideration that the types of benefits achieved will vary considerably among companies and that information requests cannot be overly prescriptive. They also need to be realistic and ensure that reporting does not place unnecessary burdens on industry. To avoid this, clear guidance materials and alignment with other reporting systems (e.g. environmental reporting, reporting on energy consumption) is warranted.

Quantifying industrial benefits

Energy efficiency in industry is particularly complex, as many processes involve multiple steps and have different energy needs along the process chain. Identifying the types of benefits that can occur from energy efficiency projects is an important step in improving understanding of the dynamics at work. If industrial sector multiple benefits are to be included in investment decisions or make a meaningful impact on policy choices, it is equally important to find ways to assign values.

Conventional tools to assess the financial impacts of energy efficiency tend to focus on the short term, using simple payback or rate of return. In some cases, companies may use longer-term investment analysis frameworks such as net present value or more complex methods. None of these methods typically considers benefits other than cost savings from reduced energy consumption. However, it is clear that energy cost savings are only a small part of the equation, especially when potential energy costs savings can be taken through a rebound effect (Rebound effect perspective 5).

Rebound effect perspective 5

Analysis of industrial sector impacts

The direct rebound effects in the industrial and commercial sectors have been investigated far less than rebound effects in other sectors (Jenkins, Nordhaus and Schellenberger, 2011). In the industrial sector, savings from reduced energy use can be directed towards more productive and value-adding activities. From the point of view of industrial sector goals, this is a positive outcome. It is also a positive outcome in terms of general policy objectives in the area of economic growth.

Improvements in productivity and competitiveness have the potential, however, to boost market share and enable increased production, which is likely to increase energy needs and generate a rebound effect from the point of view of potential energy savings targets.

© OECD/IEA, 2014.

Yet calculating other impacts can be relatively straightforward in some cases. The annual savings achieved through reduced maintenance costs generated by an energy efficiency upgrade of a certain piece of equipment can be derived, for example, from data on frequency and cost of service, duration and cost of shutdowns, and cost of maintenance materials. Including other benefits in investment calculations will shorten payback periods and contribute to raising the profile of energy efficiency. Experience in food industry projects, for example, shows that projects initially calculated to have a three- to four-year payback can deliver the full return on investment in just one year when multiple benefits are integrated into the overall assessment. (See the Companion Guide at the end of this publication for a fuller discussion of quantification methods and their application.)

In reality, some benefits are relatively easy to quantify and other very difficult. Finding a way to assess the level of complexity is useful when planning an approach to measure a specific impact, and can provide a better basis for decision making on which types of benefits to focus on. Initially, it may make sense to focus on those that are easy to quantify and, as experience is gained, move on to those requiring greater effort. While assessing the level of difficulty is a somewhat subjective task, various methods can be useful. A matrix combining the expected level of quantifiability (high, medium and low) and the time horizon (short or long term) on which the benefits are likely to occur has been created for this purpose, (Rasmussen, 2014).[5] Plotting benefits into this matrix can help to identify those that are easiest to quantify and most likely to deliver in a short time frame (Figure 5.1).

Figure 5.1 Matrix classifying industrial benefits in terms of quantifiability and time horizon

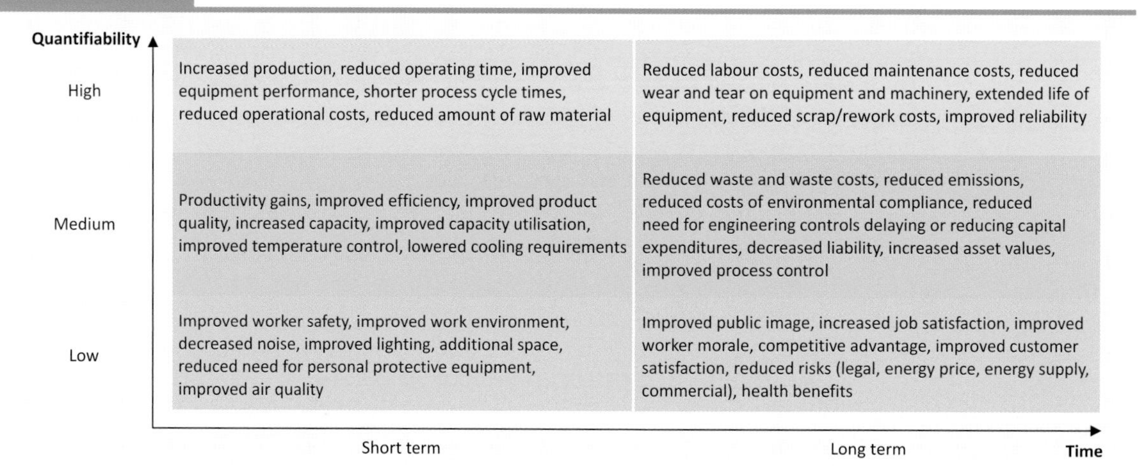

Note: Time frame is shown horizontally. Ease of quantification (termed "quantifiability"), broken into three levels (high, medium and low), is shown horizontally.
Source: Rasmussen, J. (2014), "Energy-efficiency investments and the concepts of non-energy benefits and investment behaviour", presentation at the 2014 ECEEE Industrial Summer Study, ECEEE, Arnhem, 2-5 June 2014.

Key point *Assessing first the quantifiability and time horizon of multiple benefits allows evaluators to select the best place to start early investigations, based on available resources.*

5 This approach was developed based on experiences in assessing the impacts of investments when information technology was a relatively new field (in the 1990s and the early 2000s). Categorisation was the starting point for tackling the challenge of integrating impacts into a benefit-cost framework.

© OECD/IEA, 2014.

Using this scale to assess possibility to quantify, rather than a firm division of quantifiable or unquantifiable, provides a way of dealing with the types of benefits – such as increased productivity – that fall between the two extremes. The importance of acknowledging the so-called soft, unquantifiable or intangible benefits has been stressed, at least as "extra arguments" for investment decisions (Farbey, Targett and Land, 1995; Dempsey et al., 1998).

Provided they can be quantified, there should be no methodological barrier to incorporating benefits into the variety of financial assessment methods currently used to support decision making for reporting on the benefits or impacts of energy efficiency projects. Similarly, if programmes or policies require reporting of quantified industrial benefits, this information can be used in assessing programme impacts.

Dealing with qualitative information

A remaining challenge is that some energy efficiency impacts may be less tangible, and therefore more difficult to quantify. Examples include benefits such as enhanced company reputation, improved customer loyalty or improved branding. Important issues such as avoided risk are very real factors in decision making, but also pose challenges for quantification.

Industry has developed methods to assess impacts that fall into the "low" zone on the quantifiability scale; many can be adapted to estimate values of the less tangible industrial benefits. For example, most companies are familiar with public relations or marketing operations, and the methods used to assess the effectiveness of such campaigns in changing public perception. Well-developed methodologies applied in other areas could serve as a basis for developing more robust qualitative assessment of energy efficiency outcomes. In the buildings sector, for instance, the impact of thermal comfort is sometimes assessed through effects on well-being and productivity.

Other methods include surveys of customer perceptions or the development of scenarios to show counterfactual situations and assess the likelihood of various alternatives occurring: e.g. the probability of different types of risks taking place and the associated costs for the company (or in a wider context). Superior Energy Performance (SEP), a programme sponsored by the United States (US) Department of Energy, has begun collecting qualitative data on the additional benefits reported by facilities that have received SEP certification. These data, collected through interviews with facility staff, may include information about maintenance, improved process control, reduction in waste products, increased throughput, increased productivity or other topics. Collecting this information serves the purpose of including a qualitative description of benefits that indicate the overall impact of SEP.

Estimation methods

A trade-off exists between time and effort needed to measure site-specific benefits and the exactness of results. Sophisticated quantitative calculations provide a firmer basis for decision making and/or enable comprehensive energy efficiency programme evaluations. However, developing the methods needed to make such assessments – including sourcing and verifying all information needed and conducting the calculations – requires significant time, effort and cost. Where resources are limited, using assumptions and evidence from previous experience can provide a simpler method for estimating the scope and scale of multiple benefits (see the Companion Guide at the end of this publication for more in-depth discussion).

Assessments based on site-specific data may require the use of assumptions to establish casual effects or assign values to multiple benefits. For more intangible benefits (e.g. better worker morale or better working environment) that defy direct quantification, a risk

© OECD/IEA, 2014.

exists that different stakeholders would assign different values (Lung et al., 2005). Time requirements and associated costs are a key constraint in assessing benefits; both factors may change over time as methodologies and internal routines develop, and as experience in assessing benefits increases.

Table 5.3 Choosing a method, exact calculation or estimation, according to assessment needs

Criteria	Exact calculations	Estimations
Objectivity	High/medium	Low
Time requirement	High	Low
Complexity	High	Low
Validity	Medium	Medium

Source: Adapted from Gudbjerg, E., K. Dyhr-Mikkelsen and C. Monrad Andersen (2014), "Spreading the word – An online non-energy benefit tool", presentation at 2014 ECEEE Industrial Energy Efficiency Summer Study, 2-5 June 2014, Arnhem.

Key point *A trade-off exists between time and effort needed to measure site-specific benefits and the exactness of results.*

Even in situations where it may not be possible to obtain site-specific data, it is possible to calculate the value of benefits by using information from the budget – i.e. basing calculations on how much money is spent on budget lines for the issue of interest (Box 5.9). For example, the US site of Saint-Gobain, a construction product manufacturer, implemented energy efficiency measures to optimise its compressed air system and as a result reduced water bills from installing a closed-loop system and reduced sewerage fees due to the reduction of discharge of spent cooling water to the local sewer system (Lung, 2014). Once the volumes of water savings are estimated or calculated, it is relatively easy to assign a value based on water and sewerage bills.

Making such estimations or "rule-of-thumb" assessments can be worthwhile as a first step in the process of starting to consider and assign value to the wider benefits of industrial energy efficiency.

Using multipliers

Developing multipliers[6] that could be used to indicate the order of magnitude of the value of energy efficiency benefits in the industrial sector is particularly challenging for numerous reasons. Benefits are not necessarily achieved consistently in all contexts. Moreover, the influence of several variables means that the same benefits may not be relevant each time a similar energy efficiency project is implemented. When considering multipliers in an industrial context, it is important to first identify those additional benefits that are clearly "multipliable" or transferable to other facilities. The multiplier can only be a minimum estimate of the wider benefits, and further benefits can likely be found in each specific situation.

Ultimately, assessment methods need to be transparent and easy to communicate (Box 5.10), both to practitioners engaged in assessment and to those who will use the assessment results to inform decision making.

6 The term multipliers refers to values (including default values or adders) that represent an additional value of one or a range of benefits that can then be used in lieu of an actual assessment or be multiplied with assessed energy demand reduction to derive a value.

© OECD/IEA, 2014.

Box 5.9 An example of monetisation based on existing data

Using available data to quantify and monetise some of the additional benefits of energy efficiency measures is, in some cases, quite straightforward. An evaluation of an energy efficiency measure applied to fluorescent lighting in an industrial context used the following method (Woodroof et al., 2012).

The lighting system comprised 10 000 fluorescent lighting fixtures, each with two lamps and one ballast. Each fixture consumed 60W with baseline operation of 5 000 hours per year and energy costs of USD 0.10/kWh. The baseline energy consumption was 3 000 000 kWh/yr, at an annual cost of USD 300 000.

Turning off lights when the facilities are empty reduces usage by 25%; in turn, reduced usage extends the ballast-operating lifetime by 25% and stimulates the additional benefit of reduced costs for ballasts.

A typical ballast like that used in the example has a lifespan 60 000 hours: with 5 000 hours per year operating time, ballasts need replacing at 12 years. Each ballast costs USD 20, for a total replacement cost of USD 100 000; this gives an annualised ballast replacement cost of USD 8 333. Extending the ballast lifespan reduces the annualised replacement cost to USD 6 250, delivering an annualised savings of USD 2 083 to the company.

A similar exercise can be carried out with the lamps themselves. With lamp life of 20 000 hours and a cost of USD 2.50 per lamp, reducing use by 25% has the effect of reducing annualised lamp replacement costs from USD 6 250 to USD 4 717 – delivering an annual saving of USD 1 533.

The reduced need for replacement of ballasts and lamps will also drive down associated labour costs, which also can be quantified.

Note: The additional benefit indicators calculated in the study included: reduced maintenance material; reduced maintenance labour; avoided purchase of offsets; and reduced sales taxes and environmental penalties.

Box 5.10 A pragmatic step-by-step process for early investigations

A joint project run by Lokalenergi (an electricity retail and energy service company), the Danish Technological Institute and Ea Energy Analyses (a consulting firm) shows the value of making a start on quantification methods. This team has taken the initiative to develop a *Non-Energy Benefit Tool* (NEB Tool), targeting Danish energy consultants who provide energy efficiency services primarily to industry. The underlying assumption is that easy access to information on NEBs or multiple benefits (including their scale) will lead to higher acceptance and implementation of energy efficiency projects.

The Danish research project comprises two phases. The first phase focused on developing a method for assessing NEBs, which was tested on 12 specific energy efficiency projects in close dialogue with the involved industrial and service sector companies. The second phase (started in February 2014) aims to adjust the prototype based on the feedback received from the 12 test cases and expand significantly the database of project cases.

Input of an increasing volume of energy efficiency projects should allow the team to derive valid generalisations concerning the expected type and size of NEBs associated with certain types of projects, which can be compared to international experience. This will require wide use of the database and the NEB Tool will be available online to all interested parties.

© OECD/IEA, 2014.

The NEB Tool comprises four main elements:

- a method for assessing multiple benefits of energy efficiency projects
- a searchable database, for example, searching by sub-sector and energy efficiency project type
- case examples with detailed description of energy efficiency projects and associated benefits
- a questionnaire for identifying and assessing multiple benefits.

The NEB Tool leaves it up to individual project holders to assign a value to their identified benefits based on a built-in index; this avoids having to translate all benefits into an exact monetary value. A user of the NEB Tool would go through the following process:

- Identify and classify the key benefits, and assess their relative size.
- Assess the benefit values relative to the achieved energy efficiency improvement.
- Rate the benefits relative to an index of 100.

Using the NEB Tool, an energy efficiency project that estimated increased productivity to be twice as valuable as the achieved energy cost saving, the benefit "productivity" is assigned the value +200. The individual benefits of a given project are then summarised by category and presented in a bar chart.

Source: Gudbjerg, E., K. Dyhr-Mikkelsen and C. Monrad Andersen (2014), "Spreading the word – An online non-energy benefit tool", presentation at the 2014 ECEEE Industrial Energy Efficiency Summer Study, 2-5 June 2014, Arnhem.

Policy-making considerations

In developing energy efficiency policy for the industrial sector, policy makers must adequately consider which interventions can best support the objectives and needs of diverse stakeholders within the sector. This requires identifying the nexus between the national objectives for energy efficiency (e.g. enhancing energy security, reducing carbon emissions or supporting economic growth) and the strategic objectives that drive investment decisions in industry (e.g. increasing production and developing new business opportunities). A multiple benefits approach supports this process by revealing the strategic value of energy efficiency opportunities in the business context.

At present, various barriers – including limited access to capital and technical know-how, risk aversion and up-front transaction costs – undermine the uptake of energy efficiency in industry. This means there is a potentially catalytic role for government in communicating the multiple benefits approach and encouraging stakeholders to investigate the full returns being made on energy efficiency investments. While the multiple benefits of energy efficiency in industry is a relatively new area, several recent studies (see Box 5.5 above) provide an excellent basis for further developing and piloting assessment methodologies that will be of use to policy makers and industry.

Multiple benefits of a key policy approach: EnMSs

Government can work jointly with industry to develop ways to capture industrial benefits within energy efficiency programmes. Energy efficiency improvements in industry do not always imply the need for new investments; they can also be achieved through better management of existing energy resources and energy-using systems within companies. In some industrial sectors, no- or low-cost improvements in energy management can lead to significant savings even before any investment is needed. The IEA has previously identified the significant energy savings potential offered by energy management programmes and the policies that support them.[7]

7 The IEA will also be analysing energy efficiency challenges and opportunities for small and medium-sized enterprises in a forthcoming policy pathway publication.

© OECD/IEA, 2014

EnMSs represent a collection of procedures and practices that ensure the systematic tracking, analysis and planning of energy use in individual companies. They have existed in some countries for more than 20 years and important learning has taken place during their implementation and modification. Numerous governments have designed programmes to support industry in these efforts (Box 5.11). These show that considering the wider range of benefits can benefit companies and help shape policy programme design.[8] The multiple benefits approach is expected to expand the range of positive outcomes that might be delivered by more effective energy management.

Box 5.11

Australia and Sweden develop industrial energy efficiency programmes with multiple benefits in mind

The Australian Energy Efficiency Opportunities programme requires large energy-using corporations to conduct energy efficiency assessments, and to report publicly the outcomes each year. The legislation was developed through a comprehensive industry consultation process. One concern expressed by large energy users was that evaluators would focus on the energy cost savings associated with energy efficiency projects without considering the full range of other benefits that such projects could deliver.

In response to this feedback, the government devised an evaluation method that would account for a broader range of business costs and benefits. Consequently, assessments now include (a) direct energy-related costs and savings; and (b) other quantifiable costs and benefits, including the following:

- capital cost or an avoided capital investment
- cost of maintenance, waste disposal, water usage, or occupational health and safety
- cost associated with a project delay
- cost associated with a change in productivity, or the quality or quantity of an output.

The Swedish Energy Agency is currently engaging with companies to explore the strategic potential of energy efficiency investments. The goal is to equip companies with a guide or tool that can be used to more easily quantify benefits, and thereby obtain a more complete understanding for an investment relating to energy efficiency. This approach seeks to remove some of the risk from the decision-making process.

The agency is also exploring different means of incorporating the analysis of benefits in its strategies and routines. For example, in a call for papers relating to optimising motor systems, applicants are asked to estimate benefits of their project in the application phase and again in the final project report. These requirements aim to encourage a trend of involving wider benefits in energy efficiency project decision making.

The agency also actively disseminates information regarding benefits to Swedish companies. As a result, private actors have started using this information in their communication to other parties and customers.

The boardroom perspective

As discussed in the introduction to this chapter, to develop effective policies for industry, policy makers need to understand what drives executive decision making. This is particularly important when introducing a multiple benefits approach, as a high-level commitment within the company will be needed to support increased attention to measuring the range of impacts driven by energy efficiency measures.

8 The IEA policy pathway articulates one approach to developing energy efficiency policies. That process can be developed to also appraise the multiple benefits of industrial energy efficiency measures (see Table 1 in the Companion Guide at the end of this publication).

© OECD/IEA, 2014.

The boardroom process for decision making on energy efficiency investments typically comprises five driving forces: financials, knowledge, commitment to the environment and energy efficiency, public and market demands, and policy obligations (IEA, 2011). The relative importance of each driver may vary from company to company, as may the prioritisation of benefits. While the strategic value of energy efficiency is likely to be evident in diverse industry contexts, the specificities of energy management plans will need to match the characteristics and structures of particular companies and sectors.

Policies developed through stakeholder processes and extensive consultations, especially at an early stage, can foster an increased sense of company ownership and corresponding executive support, leading to increased commitment to thorough implementation and rigorous measurement of outcomes. Similarly, using existing government consultations with industrial corporate boardrooms and other company stakeholders can provide a vehicle for discussion about the relative merits for industry of a multiple benefits approach to evaluating potential energy efficiency investments.

Further research for stakeholders

This is an area that merits significantly scaled-up research by the energy efficiency community. International co-operation and knowledge sharing are needed to accelerate progress; where policy makers engage in dialogue with industry and programme participants, such collaboration can be key to developing robust systems for assessing wider benefits of industrial energy efficiency. Several priority areas for further action have been identified in the course of this chapter (Table 5.4).

Table 5.4 Further research and collaboration opportunities in industrial sector impacts

Area	Specific actions
Benefit areas and causal linkages	Carry out a retrospective analysis to identify past policy actions and outcomes that have attempted to incorporate wider benefits; share the lessons learned from this experience.
	Strengthen research efforts and create effective linkages between research outcomes and the development of approaches or tools to enable companies, policy makers and other stakeholders to assess industrial multiple benefits and to use these assessments.
	Develop materials that could help companies, policy makers and other stakeholders to improve the way they quantify benefits; develop education/capacity building curricula and encourage more widespread adoption of good practice approaches in quantifying the wider impacts from energy efficiency projects.
Data, indicators and metrics	Implement a targeted collection of non-energy data through industry surveys, investigation of implemented energy demand reduction projects and in-depth interviews (possibly in connection to programme evaluations).
	Develop initiatives to collect country- and industry-specific case studies.
	Develop sector-specific information about types of benefits related to different energy efficiency measures and guidelines on how to assess the value of these.
	Develop costing tools that help energy managers to quantify the additional benefits of industrial energy efficiency. This could use "rule-of-thumb" estimates for various categories of benefits and energy efficiency measures, based on research and databases. Develop and implement training for energy service providers/consultants and for energy managers.

© OECD/IEA, 2014.

Area	Specific actions
Assessment methodologies	Gather and analyse data, and develop methods on how this data could be shared to create a stronger business case for including additional benefits in assessments and for developing stronger methodologies.
	Facilitate the sharing of experience and explore opportunities for developing standard methods to facilitate transfer of results.
	Integrate consideration of multiple costs and benefits in existing projects or programmes, e.g. incorporate a requirement that energy and productivity assessments include all costs related to a process or energy-using application prior to energy efficiency implementation, and that project evaluation incorporates all business benefits rather than simply energy cost savings.
	Develop and disseminate common guidelines to ensure that assessments, if not including the same factors, can at least be compared on a similar basis.
Collaboration initiatives	Convene experts with experience or interest in assessing the wider benefits of industrial energy efficiency to share experience and information.
	Create mechanisms for collaboration, such as international databases, information-sharing portals and peer-to-peer learning opportunities within industrial supply chains or particular communities of practice.

Conclusions

Including multiple benefits in the evaluations of energy efficiency programmes in industry is valuable for businesses as well as for programme providers, finance providers and policy makers. It may also have a favourable impact on cost-effectiveness calculations. From a policy-making perspective, assessing industrial benefits can enhance programme participation and help to show the broader value of the programme, which can provide strong arguments for programme continuation and help with access to funding. A better understanding of industrial benefits can also help policy makers improve programme design to better address industry needs and priorities.

The types of benefits and their value will differ significantly among different types of projects. Several challenges remain in quantifying industrial benefits, including: establishing causality, inter-linkages or overlaps among benefits; understanding direct and indirect benefits; and changes in the value of benefits over time. The overarching challenge is to assign a monetary value to these benefits so that they can be used to assess the value of projects or the results from implementation. Increased efforts by all stakeholders to collect case-by-case information on multiple benefits in industry will raise awareness of their potential value and support improved methodologies for quantifying them. As already available sources of information within industry provide a rich starting point, this does not necessarily imply new reporting requirements in connection to energy efficiency projects.

Bibliography

Cooremans, C. (2014), "Categorizing non-energy benefits of energy efficiency in strategic terms in order to boost investment", presentation at the IEA Roundtable on the Industrial Productivity and Competitiveness Benefits of Energy Efficiency, Paris, 27 January 2014, www.iea.org/media/workshops/2014/eeu/industry/1.CCooremansfinalversionCategorizingnonenergybenefitsofEEEIAJan272014.pdf (accessed 19 June 2014).

Cooremans, C. (2012), "Investment in energy efficiency: Do the characteristics of investments matter?", *Energy Efficiency*, Vol. 5, No. 4, Springer Netherlands, Houten, pp. 497-518. http://link.springer.com/article/10.1007%2Fs12053-012-9154-x (accessed 18 June 2014).

© OECD/IEA, 2014.

Dempsey, J. et al. (1998), "A hard and soft look at IT investments", *The McKinsey Quarterly*, 1998, No. 1, McKinsey and Company, pp. 126-137.

DRET (Australian Government, Department of Energy, Resources and Tourism) (2013), Case studies in systems optimisation to improve energy productivity, http://eeo.govspace.gov.au/files/2013/08/Systems-Optimisation-Case-Study-2013.pdf.

DRET (2009), *INCITEC PIVOT – Gibson Island*, Energy Efficiency Opportunities case study, DRET, Canberra, ACT, http://eeo.govspace.gov.au/files/2012/11/Incitec-Pivot-case-study.pdf (accessed 18 June 2014).

Farbey, B., D. Targett and F. Land (1995), "Evaluating business information systems: Reflections on an empirical study", *Information Systems Journal*, October 1995, Vol. 5, Issue 4, pp. 235-252, John Wiley and Sons, Inc., http://onlinelibrary.wiley.com/doi/10.1111/j.1365-2575.1995.tb00097.x/abstract (accessed 28 July 2014).

Finman, H. and J. Laitner (2002), *Industry, Energy Efficiency, and Productivity Improvements*, US Environmental Protection Agency (US EPA) White Paper, US EPA, Washington DC.

Gudbjerg, E., K. Dyhr-Mikkelsen and C. Monrad Andersen (2014), "Spreading the word – An online non-energy benefit tool", presentation at the 2014 ECEEE (European Council for an Energy Efficient Economy) Industrial Energy Efficiency Summer Study, 2-5 June 2014, Arnhem, ECEEE website, http://proceedings.eceee.org/vispanel.php?event=4 (accessed 19 June 2014).

Gudbjerg, E. (2012), Inputs provided.

Hall, N. P. and Roth, J. A. (2003). "Non-energy benefits from commercial and industrial energy efficiency programs: Energy efficiency may not be the best story", Proceedings of the 2003 IEPEC (International Energy Program Evaluation) Conference, Seattle, 19-22 August 2003, IEPEC, Wisconsin, pp. 689-702, www.iepec.org/conf-docs/papers/2003PapersTOC/papers/077.pdf (accessed 19 June 2014).

IEA (International Energy Agency) (2012), *Policy Pathway: Energy Management Programmes for Industry: Gaining through Saving*, IEA/OECD, Paris.

IEA (2011), *The Boardroom Perspective: How Does Energy Efficiency Policy Influence Decision Making in Industry?*, OECD/IEA, Paris, www.iea.org/publications/freepublications/publication/Boardroom_perspective.pdf (accessed 23 June 2013).

Jenkins, J., T. Nordhaus and M. Schellenberger (2011), *Energy Emergence: Rebound & Backfire as Emergent Phenomena*, Breakthrough Institute, Oakland.

Lilly, P. and D. Pearson (1999), "Determining the full value of industrial efficiency programs", *Proceedings from the 1999 ACEEE Summer Study on Energy Efficiency in Industry*, Saratoga Springs, June 1999, American Council for an Energy Efficient Economy, Washington DC, pp. 349-362, www.seattle.gov/light/Conserve/Reports/paper_7.pdf (accessed 22 June 2014).

Lung, R.B. (2014), "Capturing co-benefits of industrial energy efficiency in U.S. DOE programs", presentation at the IEA Roundtable on the Industrial Productivity and Competitiveness Benefits of Energy Efficiency, Paris, 27 January 2014, www.iea.org/workshop/industrialproductivityandcompetitivebenefits.html (accessed 22 June 2014).

Lung, R.B. et al. (2005), "Ancillary savings and production benefits in the evaluation of industrial energy efficiency measures", *Proceedings of the 2005 ACEEE Summer Study on Energy Efficiency in Industry*, Vol. 6, West Point, 19-22 July 2014, ACEEE (American Council for an Energy-Efficient Economy), Washington DC, pp. 6-103-6-114, www.aceee.org/files/proceedings/2005/data/index.htm (accessed 22 June 2014).

McKane, A. (2014), Inputs provided.

McKinsey Global Institute (2007), "Industrial sector", *Curbing Global Energy Demand Growth: The Energy Productivity Opportunity*, McKinsey & Company, New York, www.mckinsey.com/insights/energy_resources_materials/curbing_global_energy_demand_growth (accessed 22 June 2014).

Mundaca, L. (2014), Inputs provided.

Mundaca, L. et al. (2010), "Evaluating Energy Efficiency Policies with Energy-Economy Models", *Annual Review of Environment and Resources*, Vol. 35, Annual Reviews, Palo Alto, pp. 305-244, www.annualreviews.org/doi/abs/10.1146/annurev-environ-052810-164840?journalCode=energy (accessed 22 June 2014).

© OECD/IEA, 2014.

NAM (National Association of Manufacturers) (2005), *Efficiency and Innovation in U.S. Manufacturing Energy Use*, NAM/The Manufacturing Institute, Washington DC, www.energyvortex.com/files/NAM.pdf (accessed 22 June 2014).

Nehler, T., et al. (2014), "Including non-energy benefits in investment calculations in industry – empirical findings from Sweden", presentation at the 2014 ECEEE (European Council for an Energy Efficient Economy) Industrial Summer Study, ECEEE, Arnhem, 2-5 June 2014, www.eceee.org/industry (accessed 22 June 2014).

Pearson, D. and L.A. Skumatz (2002), "Non-energy benefits including productivity, liability, tenant satisfaction, and others – what participant surveys tell us about designing and marketing commercial programs", *Proceedings of the 2002 Summer Study on Energy Efficiency in Buildings*, Vol. 4, ACEEE (American Council for an Energy Efficient Economy), Washington DC, pp. 4.289-4.302, www.aceee.org/files/proceedings/2002/data/index.htm (accessed 23 June 2014).

Pye, M. and A. McKane (2000), "Making a stronger case for Industrial energy efficiency by quantifying non-energy benefits", *Resources, Conservation and Recycling*, Vol. 28, No. 3-4, Elseviers B.V., Amsterdam, pp. 171-183, www.sciencedirect.com/science/article/pii/S0921344999000427 (accessed 23 June 2014).

Rasmussen, J. (2014), "Energy-efficiency investments and the concepts of non-energy benefits and investment behaviour", presentation at the 2014 ECEEE (European Council for an Energy Efficient Economy) Industrial Summer Study, ECEEE, Arnhem, 2-5 June 2014, www.eceee.org/industry (accessed 22 June 2014).

UNIDO (United Nations Industrial Development Organization)/UNEP (United Nations Environment Programme) (2010a), *A Primer for Small and Medium Sized Enterprises: Enterprise Level Indicators for Resource Productivity and Pollution Intensity*, UNIDO/UNEP, Vienna, www.unido.org/index.php?id=1001348 (accessed 23 June 2014).

UNIDO/UNEP (2010b), *Enterprise Benefits from Resource Efficient and Cleaner Production*, UNIDO/UNEP, Vienna, www.unido.org/fileadmin/user_media/Services/Environmental_Management/Cleaner_Production/RECP_Peru.pdf (accessed 23 June 2014).

Woodroof, E.A. et al. (2012), "Energy conservation also yields: capital, operations, recognition and environmental benefits", *Energy Engineering*, Vol. 109, No. 5, Taylor & Francis, pp. 7–26, www.tandfonline.com/doi/abs/10.1080/01998595.2012.10531820?tab=permissions#tabModule (accessed 28 June 2014).

Worrell, E. (2014), Inputs provided.

Worrell, E. et al. (2009), "Industrial energy efficiency and climate change mitigation", *Energy Efficiency*, Vol. 2, No. 2, Springer Netherlands, Houten, pp. 109-123, http://link.springer.com/article/10.1007%2Fs12053-008-9032-8 (accessed 23 June 2014).

Worrell, E. et al. (2003), "Productivity benefits of industrial energy efficiency measures", *Energy*, Vol. 28, No. 11, Elsevier Ltd., Amsterdam, pp. 1081-1098, www.sciencedirect.com/science/article/pii/S0360544203000914 (accessed 23 June 2014).

© OECD/IEA, 2014.

Chapter 6

Energy delivery impacts of energy efficiency

Key points

- *Faced with the prospects of reduced demand from energy efficiency, energy providers are shifting to a new paradigm. The traditional business model of maximising profits by selling more units of energy is being replaced by one that recognises the opportunity in becoming a provider of energy services – including delivering multiple benefits to customers through improved energy efficiency.*
- *Energy efficiency replaces lower unit sales with direct benefits to the energy provider through avoided costs for energy generation, transmission and distribution (T&D) capacity, and line losses. Where regulation limits emissions and sets renewable resource obligations, energy providers also benefit from avoided carbon dioxide (CO_2) emissions costs and can meet obligations at a lower cost.*
- *Direct benefits to customers, such as reduced fuel and maintenance costs or improved quality and affordability of energy services, can generate additional indirect benefits for energy providers, particularly through reduced customer arrears and related management costs.*
- *Energy efficiency can bring down energy prices on the wholesale market, a benefit to all energy consumers in the market.*
- *Strong experience with energy efficiency obligations (EEOs) in the United States (US) has prompted evaluators to adapt traditional cost-effectiveness assessments to better capture the full range of energy efficiency benefits for all stakeholders in the energy delivering transaction.*
- *An evaluation of the annual impacts of energy efficiency measures carried out by energy providers in one US state incorporated a broad range of multiple benefits and found an overall benefit-cost ratio of 2.3:1 for the services offered. Incorporating water, fossil fuel and electricity savings into the analysis boosted the ratio to 2.9:1.*

Introduction: A strong evidence base for energy delivery impacts

Well-targeted energy efficiency interventions can deliver tangible benefits along the entire energy supply chain, both to power utilities and other energy providers and to end users, i.e. multiple benefits.[1] For utilities and energy providers, which are the focus of this chapter, energy efficiency can help to improve system reliability, enhance capacity adequacy, better

1 In other literature, these impacts have been variously labelled "co-benefits", "ancillary benefits" and "non-energy benefits" (NEBs) – terms often used interchangeably with "multiple benefits". The International Energy Agency (IEA) uses the term multiple benefits, which is broad enough to reflect the heterogeneous nature of outcomes and to avoid pre-emptive prioritisation of various benefits; different benefits will be of interest to different stakeholders.

© OECD/IEA, 2014.

manage peak demand, optimise utilisation of generation and network assets, create opportunities to defer generation and network investment, and dampen price volatility in wholesale markets. Moreover, energy providers can derive indirect benefits from the effect of improved energy efficiency in end-use by their customers: lower fuel costs for customers can, in turn, reduce arrears and unpaid debts, lower collection costs and, in competitive retail markets, boost customer retention. While end-user benefits are not the focus of this chapter, it is important to note that many energy provider benefits (such as improving system efficiency, which may ultimately lower prices) will generate an indirect benefit for all energy users.

A key challenge for energy efficiency in this sector is that the traditional energy provider business model is based on recuperating fixed costs by selling units of energy. This model incentivises selling more energy and obscures the business opportunity that exists for energy providers in the implementation of energy efficiency measures. A shift in paradigm is needed – and is underway – to encourage energy providers to value energy efficiency more highly.

This chapter demonstrates that energy providers can in fact benefit from energy efficiency in a variety of ways. In addition to providing an overview of the multiple benefits of energy efficiency to energy providers, it highlights the methodologies used to measure them. It also outlines the challenges that some broader benefits can present for measurement and lessons learned to date in attempts to quantify the "hard-to-measure" benefits. A better understanding of the added value of energy efficiency programmes can help both energy providers and policy makers to better evaluate the potential opportunities to achieve economic and social objectives through energy efficiency activities, while also achieving benefits for providers.

The EEO context

To date, most energy efficiency activities involving energy providers have been driven by government-imposed obligations; this is especially true among providers with customers that have small-scale energy demands. Referred to as EEOs or white certificates,[2] these policies have proven successful in both regulated and liberalised markets (Box 6.1). In the United States, where there is a long history of regulated EEOs, energy utilities have significant experience in measuring the attainment of EEO targets and examining, over time, the multiple benefits arising from EEOs. Consequently, much of this chapter focuses on US-based evidence of the benefits to energy providers.

Growing recognition of the multiple benefits arising from EEOs may have contributed to the paradigm shift now underway among some energy providers, reflected in a move away from simply managing energy *sales* to a new focus on providing energy *services*, which includes supporting the efficient use of energy. The implications of this will vary depending on the market conditions and regulatory environment in which a given energy provider is operating (Box 6.2).

2 EEOs generally require the energy provider to meet a clearly defined energy saving target within a certain time frame, and to report on results. The obligated energy provider is awarded a white certificate for verified energy savings. The EEO term is generically used here to cover both traditional EEOs and the few jurisdictions in which open generation and trading of white certificates is possible (e.g. Italy and two Australian states).

© OECD/IEA, 2014.

Box 6.1 Good prospects for EEOs in both vertically integrated and liberalised markets

There is a strong history in using EEOs as a policy instrument, particularly in regulated, vertically integrated markets where implementation and monitoring of targets is easier due to the smaller number of participants: the obligations often involve only a single regulator and a single utility.

EEO benefits are potentially available to energy providers in liberalised markets as well, as evidenced more recently in Europe. Indeed, many earlier European EEOs were established within a regulated system and then made the transition to operating in increasingly liberalised energy markets, which are typically more complex and require a regulatory framework that equally shares costs and benefits across the value chain and among all stakeholders.

Liberalised markets tend to make data more available and are thus more transparent, which makes it easier to capture and monitor different facets of energy efficiency. The competitive environment of a liberalised market should also promote innovation and least-cost solutions for energy efficiency. Some European retailers are now developing energy service divisions to take advantage of the new business opportunities presented by energy efficiency.

The United States has taken a lead role in multiple benefits analysis, recognising the need to account for the full benefits of energy efficiency from a cost-effectiveness perspective. In fact, the pioneering *California Standard Practice Manual* (CA SPM)[3] has been widely used for many years as a guide on how to apply various cost-effectiveness tests to energy efficiency initiatives.[4] More recently, practice has advanced to encompass several key challenges current energy regulators face, which the CA SPM originally did not address. Analysis has expanded to cover a broader range of the identified multiple benefits, beginning to quantify their diverse values (NESP, 2014; Synapse Energy Economics, Inc., 2013).

Box 6.2 Reducing energy sales still makes sense for energy providers

The idea that energy suppliers should seek to reduce energy demand can appear counter-intuitive, as it means they ultimately distribute and sell a lower volume of their main product.

In fact, an energy provider business model that incorporates energy-savings activities is becoming commonplace, as demand for energy grows globally while concern for energy security is rising. This is particularly true in countries and regions with competitive and open energy markets. Providing energy efficiency services to customers represents a new avenue of business for energy providers (in addition to unit sales of energy).

In order to avoid a perverse impact on the business case for energy providers, most EEOs include a mechanism to ensure that energy providers are not disadvantaged by reduced distribution or sales. This mechanism, known as "decoupling" in North America or "removing the volume driver" in Europe, usually compensates the utility for its investment in energy efficiency activities and for lost revenues. Revenue adjustment mechanisms or partial decoupling, a further refined mechanism, promote energy efficiency directly by additionally compensating utilities for their energy efficiency activities. Increasingly, the value of multiple benefits shows potential to reduce the need to compensate energy providers for their energy efficiency activities, although calculating a monetary value for these benefits remains a key challenge. Many US states offer a form of incentive regulation to financially reward the best performers in EEOs.

3 This guide was first introduced in the 1980s and has been revised and updated several times, most recently in 2001 (http://cleanefficientenergy.org/resource/california-standard-practice-manual-economic-analysis-demand-side-programs-and-projects.

4 Cost-effectiveness tests are also known as "screening tests".

© OECD/IEA, 2014.

The range of energy delivery impacts

Energy efficiency measures can be implemented throughout the energy delivery value chain, from generation to T&D (i.e. in the supply side), and in final consumption by energy end-users (i.e. in the demand side). While energy utilities tend to invest in measures at all stages of the process (e.g. improved generation technologies) in order to improve operational efficiency and reduce fuel needs, this chapter focuses on demand-side energy efficiency measures and their system-wide impacts.

Energy efficiency in this chapter refers to interventions that aim to reduce energy consumption or demand as well as those that target load reduction and load shifting. These programmes can include providing advice to customers on relevant energy efficiency measures, helping them to access financial incentives, and direct installation (equipment replacement); they can also be based on bulk procurement and/or distribution. The recent but rapid penetration of information and communication technology (ICT) tools on the end-user side support this shift for both energy providers and their customers: consumers can more actively control their energy use while energy providers can better monitor, aggregate and control end-use loads.

The multiple benefits that arise from the energy provider perspective can be divided into two broad categories:[5]

- **direct benefits** to the energy provider or to the energy customer
- **indirect benefits** which can be further broken down into two sub-categories:
  - flow-on benefits arising from direct benefits that accrue to the energy customer
  - benefits for all consumers in an energy market.[6]

Table 6.1 Energy provider multiple benefits arising from energy efficiency

Direct benefits	Indirect benefits
Avoided transmission capacity costs	Reduced credit and collection costs
Avoided generation operation costs	Reduced financial risk
Avoided CO_2 costs	Anticipating future environmental regulation costs
Avoided other environmental regulations costs	Improved customer retention
Avoided line losses	Improved corporate relations
Avoided generation capacity costs	Reduced maintenance costs*
Avoided transmission capacity costs	Saving of other fuels*
Minimising reserve requirements	Reduced water usage*
Reduced cost of Renewable Resource Obligation	Employee productivity improvements*
	Increased comfort (e.g. health and well-being)*
	Increased health and safety*
	Reduced prices in wholesale market**

* This is a direct benefit to customers.
** This also benefits all customers.
Source: Unless otherwise noted, all material in figures and tables in this chapter derives from IEA data and analysis.

5 Other categorisations are possible; e.g. the categorisation commonly used for "non-energy benefits" in North America breaks benefits into three categories: utility, societal and participant (Skumatz, 1997).

6 This chapter considers multiple benefits for end-users only from the perspective of their role as customers of the energy provider. Additional impacts for society, such as environmental, health and macroeconomic benefits arising from energy efficiency, are addressed in more depth in the preceding chapters.

© OECD/IEA, 2014.

Ultimately, most multiple benefits to the energy provider will also benefit their customers and society at large. For example, in a vertically integrated environment, benefits from energy efficiency activities will likely be taken into account when an energy provider faces future price reviews; if the regulator decides to reduce the price to reflect a positive energy efficiency effect then it would benefit all customers, not just those who participated in the EEOs administered by their own energy provider. In a competitive market, energy retailers who deliver such obligations more cost-effectively than their competitors will have a price advantage that can be passed on to customers. Their energy efficiency activities may attract new customers or better retain existing customers. A first step towards that goal is to identify the full range of benefits – both direct and indirect (Table 6.1).

Direct benefits to energy providers

The most significant direct benefits are avoided energy generation costs, distribution upgrade deferment and reduced line losses. Typically, these three benefits add up to around 80% of total energy provider benefits. Lower generation also means energy providers have lower line losses and lower reserve requirements. When energy efficiency dampens demand growth, energy providers may be able to reduce, delay or defer investments in expanding T&D infrastructure. Direct benefits that are more difficult to quantify but also important include risk mitigation (e.g. environmental regulation) and potential losses associated with customer management.

The impact assessment carried out by Efficiency Vermont, a US energy regulator, provides an interesting case study (the "Vermont study") (Efficiency Vermont, 2012) that demonstrates how benefit values for energy efficiency, to both energy providers and energy customers, can be calculated (Box 6.3). The results are referred to throughout this chapter.

Understanding these benefits requires detailed knowledge of two things: system load dynamics throughout the year; and the value of the system elements affected by the energy efficiency activity. The system dynamics and system elements engaged in meeting peak demand are particularly important – and should be found in data held by individual energy providers or system operators.

Box 6.3 Efficiency Vermont: An extensive assessment of energy provider multiple benefits

Efficiency Vermont is the entity that acts as the EEO scheme administrator operating under the Vermont Public Service Board in Vermont, United States. It provides technical support, as well as rebates and other financial incentives, to Vermont households and businesses to reduce their energy consumption, often through installation and retrofitting of energy efficient technologies and infrastructure. A strong emphasis on the multiple benefits of energy efficiency as a key driver of the state's economic development and environmental health has made the Efficiency Vermont experience a benchmark in this emerging approach.

In its 2010 Annual Report (the "Vermont study") (Efficiency Vermont, 2012), Efficiency Vermont demonstrates that the energy efficiency programmes reduced energy demand by 110.8 gigawatt hours over the 10.4 year average life of the measures – at a total cost of USD 33.5 million. This results in a levellised cost of energy efficiency measures of USD 39 per magawatt hour (USD/MWh. In evaluating the annual impacts of energy efficiency measures, the Vermont Study takes into account a comprehensive range of the multiple benefits to energy providers (see Figure 6.1, Table 6.2 and Table 6.6 below). This approach found an overall benefit-cost ratio of 2.3:1 for the services offered.

© OECD/IEA, 2014.

To determine the quantified value of each benefit, the Vermont study calculated the NPV of savings in each benefit area, levelised over the average duration of the measures installed in that year (10.4 years). This provided an annual value, which was then divided by the annualised energy savings of 110 800 megawatt hours to yield a dollar value per megawatt hour. Ultimately, these energy provider multiple benefits (together valued at more than USD 104.8/MWh) should benefit all customers (see discussion below on how the additional benefits accruing to participating end users boost the overall value).

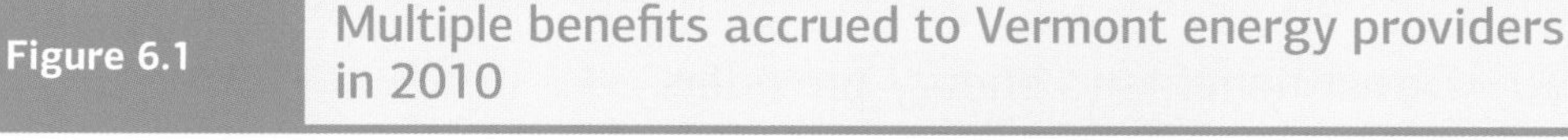

Figure 6.1 Multiple benefits accrued to Vermont energy providers in 2010

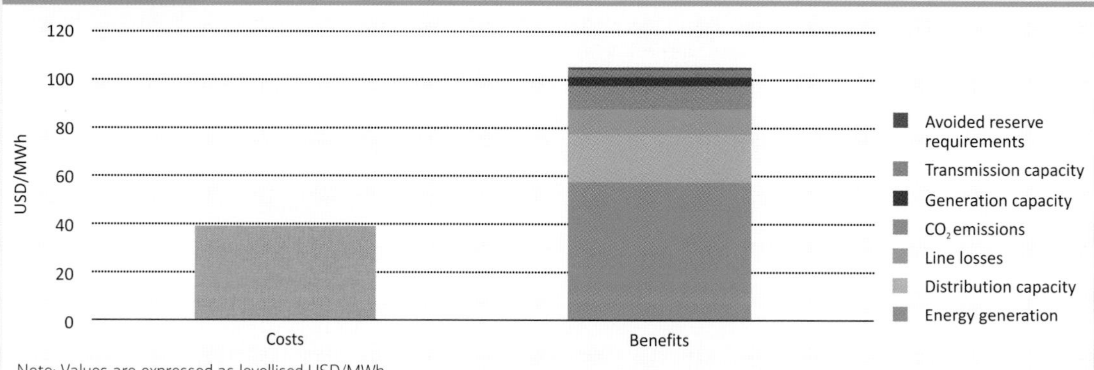

Note: Values are expressed as levellised USD/MWh.

Source: Efficiency Vermont (2012), *Annual Report 2010*, Efficiency Vermont, Burlington, www.efficiencyvermont.com/docs/about_efficiency_vermont/annual_reports/2010_Annual_Report.pdf.

Table 6.2 Multiple benefits accrued to Vermont energy providers in 2010

Benefit to energy provider	Results (USD/MWh)*
Avoided existing environmental regulation costs (not CO_2)	Small
Reduced cost of Renewable Resource Obligation	0**
Minimising reserve requirements	0.7
Avoided transmission capacity costs	3.2
Avoided generation capacity costs	3.8
Avoided CO_2 emission costs at USD 20 per tonne	9.4
Avoided line losses	10.2
Avoided distribution capacity costs	20.0
Avoided generation energy costs	57.5
Sub-total of all energy provider multiple benefits	**> 104.8**

Note: For comparison the annual average retail price for electricity in Vermont in 2010 was USD 132/MWh and the average wholesale price in the regional Market New England independent system operator was USD 50/MWh (US EIA, 2012).

* All benefits are levellised in USD 2010 currency rates.

** Vermont does not have a binding renewable obligation target expressed in percentage of electricity provided, but the value for other northeastern States has been estimated at USD 1.8/MWh to USD 6.3/MWh.

© OECD/IEA, 2014.

Energy generation cost savings

Energy costs include the costs of fuel and operating expenses associated with the production of electricity. These costs are dictated by the specific operating time and plant type of the marginal generation unit. A reduction in electricity usage delivered through energy efficiency measures can directly avoid these costs. Avoided energy generation costs refers to those costs a utility would, in the case of no demand reduction, have had to incur to generate electricity (or purchase it from another source) to balance supply and demand; in other words, the portion of energy saved and the associated cost savings. Avoided costs are calculated by dividing total financial savings by the total megawatt hour savings.

In the Vermont study, the net present value (NPV) savings from this energy cost reduction was reported to be USD 49 million (see Box 6.3). Levellised and annualised, this benefit was valued at USD 57.5/MWh, more than half of total benefits (USD 104.8/MWh). Assessment of a recent EEO in the United Kingdom shows the value of comparing possible scenarios (i.e. implementation of the energy efficiency measures against a business-as-usual [BAU] case) in advance to better understand the range, scale, scope and value of the multiple benefits (Box 6.7 below).

Avoided T&D investment costs

Many energy efficiency measures aim to reduce overall energy consumption in the homes or premises of end-use customers. The value of this, in terms of multiple benefits for energy providers, increases when the lower consumption occurs during periods of peak demand, as it reduces the need to augment the T&D system capacity to handle peak load. Location of the reduced demand is crucial in determining the distribution capacity benefits of energy efficiency (even more than production or transmission). Depending on the energy efficiency programmes and their regional targeting, reduced demand in particularly remote or densely populated areas can allow energy providers to defer (for some time or even indefinitely) the need to renew the T&D system (Neme and Sedano, 2012).

Utility marginal cost studies typically value distribution system capacity costs at USD 50 per kilowatt hour per year (USD/kWh/yr) to USD 100/kWh/yr, based on the utility forecasted distribution system upgrades planned in the five- to ten-year time horizon (Lazar and Colburn, 2013). Whereas electricity rates are based on average distribution costs (including operating expenses), energy efficiency avoids marginal distribution costs (which typically involve higher capital costs than historical average costs on which rates are computed).

In the Vermont study, regulators jointly assessed the value of avoided capacity costs for both transmission and distribution (they did not include T&D maintenance expenses). Transmission and distribution were separated using an average of the ratio of values reported by New England utilities (Lazar and Colburn, 2013). The allocated cost savings for transmission (USD 2 728 million) and distribution (USD 17 million) delivered savings of USD 3.20/MWh and USD 19.99/MWh, respectively.

Several energy providers have successfully captured network benefits from energy efficiency by targeting efforts on constrained areas where the costs of adding new generation or T&D capacity are high. Con Edison, for example, successfully mobilised its energy efficiency programmes to help relieve pockets of congestion and network overloading in New York City (Box 6.4). This service area represents a very dense electrical load, with a steep load duration curve driven by commercial customers. Toronto Hydro has similarly alleviated distribution capacity pressures by integrating its capital refurbishment programme with its conservation and demand management programme, allowing them to identify opportunities to achieve energy savings obligations while also deferring network additions

© OECD/IEA, 2014.

and managing over-loading. Assessments made by Toronto Hydro indicate that spending CAN 1 million annually on demand management can offset a CAN 80 million network addition (Tyrell, 2013).

Demand-side management is made much easier by digitally controlled ICTs available to both energy providers and end-use customers. These devices can offer tangible financial benefits including reducing the cost of providing ancillary services for grid stability, reducing congestion costs on transmission networks, and facilitating better integration of distributed generation (e.g. electric vehicles connected to the distribution system). Electricity industry estimates for the potential benefits that could be unlocked by connected homes and businesses are impressive: a recent study by the Consortium for Energy Efficiency[7] estimated a reduction of up to half of peak load, and bill savings of as much as 25% (Wisniewski, 2013). Further evaluating the early benefits delivered by these technologies is a valid undertaking to support uptake of their role in energy efficiency activities.

Box 6.4 Avoided system/network costs: An example from Con Edison in New York

Steady growth in peak demand (about 1% annually) has led to a situation in which Con Edison's New York City networks are constrained for a few hours every year. Adding capacity in this geographic location is almost prohibitively expensive, not to mention disruptive.

To address the situation, Con Edison developed planning and business models that maximise the deferral value of targeted energy efficiency and demand-side management (DSM). Integrating demand-side resources into network planning offers a hedge against demand growth and an option to defer expansion projects until they are really needed. Of course, this has to be done without introducing undue risk from weather-related demand spikes and overloads.

The planning and business models involved creating load duration curves for each network, identifying the localised impacts of energy efficiency and demand-side investments, and occasional targeting of programmes to specific assets.

Each of Con Edison's 84 underground networks is independent. Thus, it is possible to project individual network peaks by disaggregating system-wide growth, considering specific load additions in each network and weather-adjusting the results.

Once a constrained network is identified, Con Edison sets a demand reduction target and procures DSM services through a reverse auction. Third-party DSM providers bid in to offer a demand reduction in that neighbourhood over a set time period. Con Edison's experience shows that specific DSM programme offerings can be designed to suit specific network constraints.

Through localised energy efficiency and DSM programmes, Con Edison projects it can limit demand growth on constrained networks to 1.2% annually between 2012 and 2021, down by 0.4% compared to the baseline case without DSM. Financial analysis estimates the network benefits of reduced demand growth through DSM at USD 1 billion annually, on an annual investment of USD 400 million for the energy efficiency and DSM programme.

Forecasting and localised planning is key to realising these network deferral benefits. Con Edison's planning guidelines now call for planners to look first at customer resources before considering capacity additions.

A similar approach is being used to target legacy consumers of Con Edison's decades-old steam system. The company offers incentives to upgrade steam-fired heating and cooling systems, thus avoiding adding new electrical demand.

Source: Craft, R. (2012), *Network Deferral Benefits of Energy Efficiency*, presentation at the PEPDEE North American Regional Energy Policy Dialogue, IEA and RAP, Washington DC, 18-19 April.

7 The Consortium for Energy Efficiency (CEE) is composed of investor-owned or municipal utilities and government, state or provincial energy offices and agencies from Canada and the United States.

© OECD/IEA, 2014.

Avoided cost of line losses

Line losses are due to the distance power is conveyed and the resistance of the wires carrying the electricity from the generator to the end-user. In transmission and distribution systems of member countries of the Organisaton for Economic Co-operation and Development, the typical utility's average resistive losses are about 7% over the course of the year and about 11% (i.e. 1.5 times higher) during extreme peak. These losses grow exponentially with system load and are proportional to the square of the current in the wires.[8]

Energy efficiency programmes reduce peak load and therefore the peak capacity that needs to be transported through the T&D lines. The marginal losses, those that arise if the load demand increases or decreases by a unit, are higher than the average losses due to the exponential relationship between losses and current. At the extreme peak, the marginal resistive losses are therefore much higher (20% in Figure 6.2) compared to the average losses (11% in Figure 6.2). This means a significant benefit derives from measures that reduce peak demand, including energy efficiency, demand response and use of emergency generators located at customer premises. Providing an additional 1 kilowatt (kW) load at peak actually requires 1.25 kW of generation, if marginal line losses of 20% are considered (Figure 6.2). This benefit from reduced line losses represents both energy and capacity (including T&D capacity) cost reductions, as direct results of electricity savings from energy efficiency measures. In the Vermont study, the NPV total savings from reduced line losses was estimated to be USD 8.7 million, giving a levellised value of USD 10.2/MWh.

Figure 6.2 Comparison of average and marginal line losses for typical US utility as a function of the load

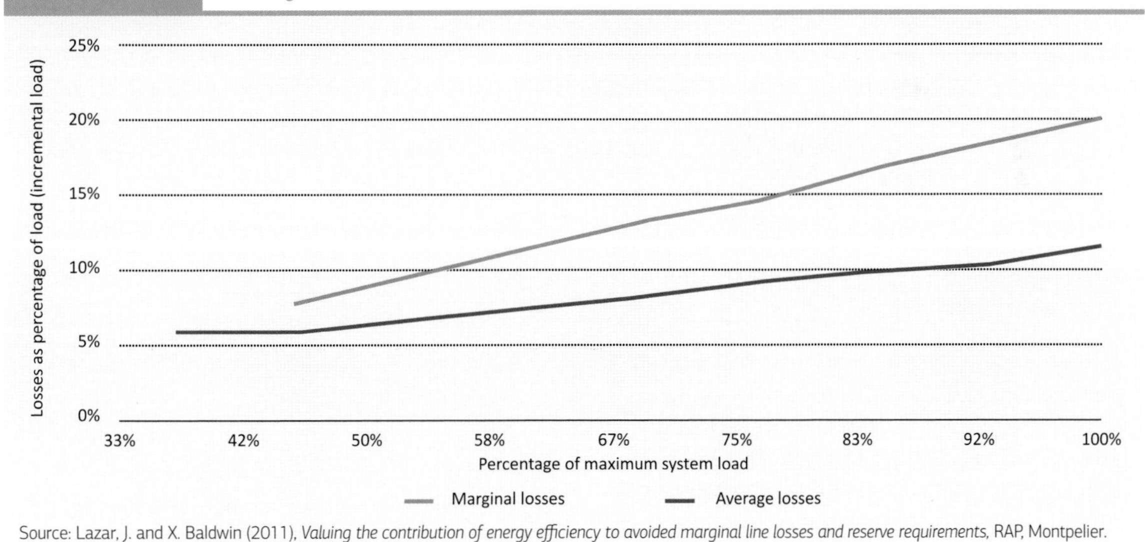

Source: Lazar, J. and X. Baldwin (2011), *Valuing the contribution of energy efficiency to avoided marginal line losses and reserve requirements*, RAP, Montpelier.

Generation capacity investment savings

Generation capacity is designed to meet annual peak demand at all times during a year. A reduction in annual peak thus results in generation capacity savings. The methodologies for covering capital costs vary in practice and depend on the market. Some electricity systems, such as the major PJM system in the northeastern United States, have explicit forward capacity markets to ensure adequate capacity to meet peak load. In such markets,

8 These losses grow exponentially with system load and are proportional to the square of the current in the wires.

© OECD/IEA, 2014.

electric capacity cost savings can be viewed as the reduction in the annual forward generation capacity to meet peak demand generation.

Explicitly quantifying the benefit involves identifying the peak savings from energy efficiency measures and their electricity load shape. To do this, the New England Independent System Operator took the approach of dividing the data for the New England capacity market into three time periods – peak, intermediate and all other hours – each with different capacity costs in USD per megawatt hour. As many energy efficiency measures correlate with peak demand (e.g. lighting and space/water heating in winter; air conditioning and commercial lighting in summer), prices and policy options can be developed for these critical times. The saving calculated for reduced production capacity in Vermont in 2010 was USD 3.23 million, levellised and annualised to USD 3.79/MWh.[9] As comparison, this represents roughly 10% of wholesale price in New England ISO, the regional market in which Vermont participated.

Another benefit of energy efficiency that results from lower peak demand is lower reserve capacity requirements for generation. Reserve capacity, generally in the range of 10% to 15% of total generation capacity, is needed to cover unexpected or planned outages. Energy efficiency can reduce the need for the utility to purchase or maintain that additional reserve capacity. In Vermont, the value of capacity reserves was assumed to be 15% of the total dollar value for avoided energy generation capacity, giving a value of USD 0.67/MWh.

Integrated system analysis

Ultimately, the electricity system is most efficient when it operates as an integrated system, i.e. when the various constituent parts (presented separately above for ease of analysis) function in concert. The benefits of energy efficiency will be felt across the system, in various ways and in varying manners, depending on the characteristics of the system, of the demand profile, of the energy efficiency interventions and other factors. Assessment of the integrated system can provide insights to drive the design of energy efficiency programmes (Box 6.5). The challenge is to develop markets and regulation that enable the sharing of costs and benefits across the system to achieve system-wide benefits.

Box 6.5 Germany: Quantifying system-level direct benefits

A recent study offers the first quantified look at the electric system benefits of deep investment in energy efficiency in Germany (Wünsch et al., 2014). The study compares a BAU (baseline) scenario with three others to examine varying reductions in electricity consumption:

- Energy Plus with 16% reduction compared to today's consumption on a long-term basis;
- Energy Concept with 20% reduction; and
- World Wide Fund for Nature (WWF) scenario with 40% reduction.

All scenarios model the German government's commitment to renewable electricity sources dominating the power system by 2050, having a share of 81% (Prognos AG, EWI and GWS, 2010). The scenarios optimise use of load management to achieve a decrease in load peaks and further expand electricity storage systems. Costs for the four scenarios for 2035 are shown in Table 6.3.

9 This value is highly dependent on capacity shortage. The recent recession initially reduced demand for new capacity but retirement of some old power plants is leading to a need for new capacity in the New England and PJM capacity markets.

© OECD/IEA, 2014.

Table 6.3 Total annual costs for German power generation and grid infrastructure in 2035

Scenario	BAU	Energy Plus	Energy Concept	WWF
2035 EUR billion (in 2012 currency)	65	55	52	44

The energy system benefits assessed include T&D benefits, and offset demand for renewables and offshore grid interconnections. Each scenario shows significant cost reductions from improved energy efficiency, with the WWF scenario showing that power system costs can drop on a mid- to long-term basis, even with greater use of renewable energy.

Costs per megawatt hour of generated electricity and grid infrastructure are about the same in all scenarios, at EUR 120 (2012 money). The cost of renewable energy is the biggest element in the scenario estimates, but can be significantly minimised by improving efficiency. In fact, energy efficiency will represent more than half of savings by 2035 in the three energy efficiency scenarios. The value of these savings, in levellised costs, is in the range of USD 150/MWh to USD 205/MWh (EUR 110 per megawatt hour [EUR/MWh] to EUR 150/MWh). The study highlights that comprehensive, long-term and aggressive investment in end-use energy efficiency in Germany will yield large power sector cost savings.

Figure 6.3 Potential electricity systems savings in Germany: 2035

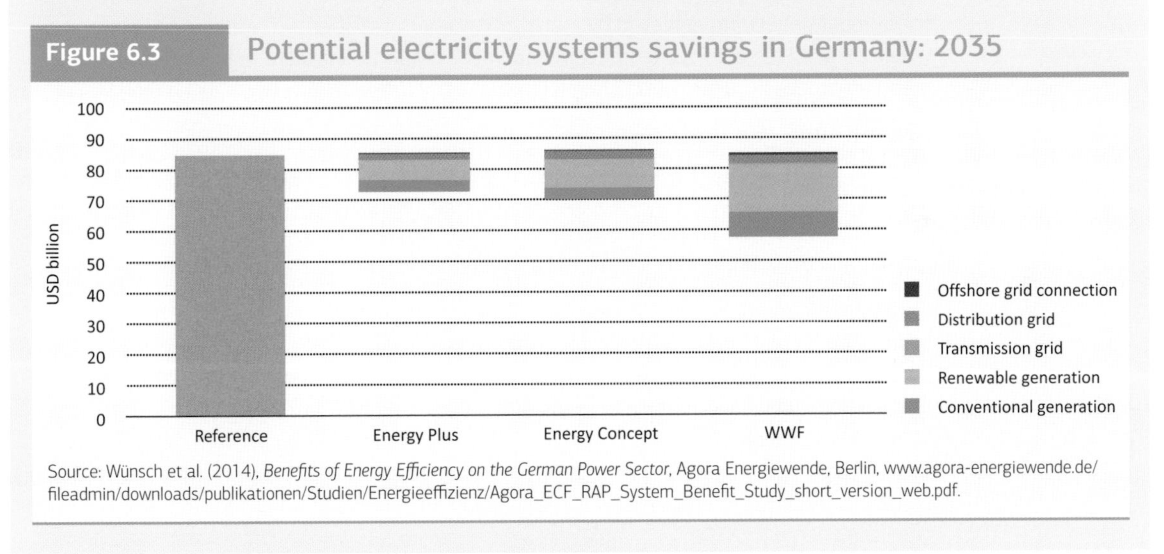

Source: Wünsch et al. (2014), *Benefits of Energy Efficiency on the German Power Sector*, Agora Energiewende, Berlin, www.agora-energiewende.de/fileadmin/downloads/publikationen/Studien/Energieeffizienz/Agora_ECF_RAP_System_Benefit_Study_short_version_web.pdf.

Local environmental regulation

Fossil fuel-based electricity generation is known to produce emissions that have detrimental effects for the health of people and the environment, including sulphur dioxide (SO_2), mercury, particulates and nitrogen oxides (NOx). Rates of these emissions have been reduced over time on a global scale through regulation and the creation of markets for the more important environmental emission allowances (e.g. SO_2 and NOx). This reduction has been further accelerated by the increased penetration of natural gas as feedstock for electricity generation and through technological improvements. As a result, the financial benefit to the energy provider (not the estimate of the costs arising from the resulting damage) of further reducing these emissions through energy efficiency has decreased over time on a megawatt hour basis. In the United Kingdom, the air quality benefits have been valued at less than 1/10th the value of the energy resource saving benefits. However,

© OECD/IEA, 2014.

improving air quality remains an important issue (particularly in industrialised regions such as Krakow, Poland) and is becoming more urgent in many emerging economies (such as China).

Quantifying the emission reduction involves a straightforward calculation of the average SO_2 or NOx content per megawatt hour in relation to the real or net energy savings achieved by energy efficiency. This emission saving is then valued at the national or regional price per tonne of avoided SO_2 or NOx.

Local governments have played a key role in reducing emissions by implementing environmental regulations, which create current and future compliance liabilities for energy providers. The liabilities include capital costs and running costs for controls and monitoring, variable operations and maintenance (O&M) costs associated with equipment and activities, allowance costs (where a "cap-and-trade" programme exists), and permitting and emission fees. In many jurisdictions, these costs are currently included in the cost of electricity delivered; however, it is not clear that future environmental requirements and costs are being taken into account in the planning process.

In the future, it is anticipated that pollution control costs will increasingly be internalised in energy provider calculations, as new environmental regulations are adopted and currently externalised costs (e.g. health and other damage costs of emissions) are integrated into financial evaluations. Monetising avoided compliance costs together with monetised values for air quality benefits achieved by substituting energy efficiency for electricity generation would allow these environmental costs and benefits to be included in the integrated planning process (Colburn, 2013).

Greenhouse gas emissions

More and more governments are imposing a financial cost for carbon dioxide (CO_2) emissions associated with energy generation. Mechanisms include carbon pricing (such as an emissions trading scheme [ETS]) or carbon tax, as well as implicit pricing through regulations that target CO_2 reductions and their corresponding shadow-prices for carbon.

Carbon markets are becoming more widespread: active markets include the European Union ETS (EU-ETS) the Regional Greenhouse Gas Initiative (RGGI) operating in northeastern United States and, more recently, seven ETS projects being piloted in China at provincial and municipal levels. Although the EU-ETS market attained a price of ~USD 20 per tonne of CO_2 (/tCO_2) in 2010, current values are much lower, being ~USD 7 in the ETS and ~USD 2 in the RGGI. While prices have varied, the basic principle has remained constant: fossil fuel-based electricity generation will engender a cost for the emitter, in this case the energy provider.

Energy efficiency has the potential to lower these emissions and the related costs borne by energy providers as the emitters. Quantifying the benefit of energy efficiency in regard to reducing the cost of CO_2 emissions is relatively straightforward: the average CO_2 content of the electricity (in kilogrammes of CO_2 per kilowatt hour) is known and is simply multiplied by the energy saved. In the Vermont case, it was calculated at USD 9.40/MWh. In the United Kingdom, since 2002, the cost (from the national perspective) of saving 1 tonne of carbon dioxide (CO_2) has been one of the key quantities evaluated and used to justify further EEO activity (Box 6.6).

© OECD/IEA, 2014.

Box 6.6 Cost of saving energy and CO_2 in the United Kingdom through EEOs

The United Kingdom introduced EEOs in 1994. At the end of each phase of the EEO (typically three years), the United Kingdom (UK) government undertakes a detailed evaluation to examine EEO performance and to improve future EEOs. The evaluations are wide-ranging, but of greatest relevance in the context of assessing multiple benefits for energy providers are the cost-effectiveness and value of avoiding a unit of energy demand, and the NPV per tonne of CO_2 saved (a key policy objective).

For the EEO phase ending March 2008, the UK government reported that the cost to energy retailers of saving one unit of electricity was USD 28/MWh (GBP 17 per megawatt hour [GBP/MWh])* after correcting for comfort, net savings, heat replacement effect and free-riders. This is much less than the energy resource cost savings of USD 87/MWh (GBP 50/MWh) (UK DECC, 2008).** The national cost-effectiveness for saving natural gas was calculated at USD 38/MWh (GBP 22/MWh). After all the corrections above were taken into account, the cost to the nation of saving 1 tCO_2 was USD -77/MWh (GBP -45/MWh) – i.e. a benefit, not a cost.

The Department of Energy and Climate Change (DECC) recently carried out an impact assessment of proposed changes to the current EEO, which would address CO_2 savings, as well as energy savings, fuel poverty impacts, air quality benefits and employment compared to BAU (Table 6.4).

Table 6.4 UK EEO assessment compares BAU against a proposed Central Scenario

Benefit to energy provider	BAU (USD)	Central Scenario (USD)
Installation costs	4 930	3 690
Hidden/hassle costs	1 967	1 639
Assessment costs	716	485
Finance costs	383	419
Admin costs	362	363
Green Deal mechanism costs	444	303
Total costs	**8 807**	**6 899**
Energy savings	8 308	6 677
Comfort benefits	1 819	1 600
Air quality benefits	706	633
Lifetime non-traded carbon savings	2 212	1 312
Lifetime European Union allowance savings	238	225
Total energy provider and end-use consumer benefits	**13 287**	**10 449**
NPV	**4 481**	**3 550**
Lifetime non-traded carbon savings	24 $MtCO_2$	14 $MtCO_2$
Cost-effectiveness	- USD 95/tCO_2	- USD 160/tCO_2

Note: The UK example does not reflect all the multiple benefits to the energy delivery chain identified in this chapter (UK DECC, 2014).
* All figures use the HM Treasury recommended discount rate of 3.5% real.
** The figures on the cost of saving energy reported above are slightly higher than those in the report due to using the latest values for the heat replacement effect.
Source: UK DECC (2014), *The Future of the Energy Company Obligation*: Consultation Document, United Kingdom government, London, www.gov.uk/government/uploads/system/uploads/attachment_data/file/291900/Energy_Company_Obligation__ECO__The_Future_of_the_Energy_Company_Obligation_Consultation_DocumentFINAL.pdf (accessed 6 July 2014).

© OECD/IEA, 2014.

Reduced costs of renewable resource obligations

Benefits associated with reduced costs for renewable obligations arise when the energy provider is legally bound to using renewable energy sources to supply a certain percentage of their energy/electricity. Such obligations are commonly called Renewable Portfolio Standards (RPS) in the United States. To the extent that the price of renewable energy exceeds the market price of electricity, energy providers incur a cost to meet the obligation or RPS percentage target. To compensate for this shortfall, that incremental unit cost is recovered through the price of a Renewable Energy Certificate (REC), a subsidy that approximately equalises the renewable electricity and wholesale electricity market prices. This annual compliance cost equals the quantity of renewable energy purchased (megawatt hours), multiplied by the REC price (US dollars per megawatt hour).

This benefit was examined for the five northeastern states (Connecticut, Maine, Massachusetts, New Hampshire and Rhode Island) that had percentage RPS targets (the Vermont RPS did not have such a target). Energy efficiency programmes reduce the cost of meeting RPS requirements by reducing the total load (or megawatt hours) that the energy provider must supply. In turn, this reduces the costs the energy provider recovers from their end-use customers to comply with the RPS.

Using a regional supply curve of renewable energy potential, it is possible to calculate a levellised RPS benefit for the 2014-28 period as lying between USD 1.80/MWh and USD 6.30/MWh (in 2013 dollars) (Synapse Energy Economics, Inc., 2013).[10] Clearly, this result depends on the characteristics of the individual systems and regulatory policies; in general, the while these values are significant, they are lower than the line loss and T&D benefits. As is discussed previously with regards to environmental regulation, it is likely that the importance of this benefit will increase in the future.

Some indirect benefits linked to customers

Capturing the full range of multiple benefits that accrue to customers as a result of energy efficiency programmes undertaken by energy providers is key to calculating values in this area. Direct financial benefits (e.g. lower energy bills) to the consumer can provide indirect, but measurable, financial benefits to energy providers in the form of lower credit management costs and fewer customer bill payment delinquencies. Not all customer benefits represent a quantifiable benefit to energy providers but many can influence the energy provider's reputation and customer loyalty. Standard practice has already developed to support the inclusion of some of these benefits in programme assessment.

Improved affordability drives reduced credit and collection costs

By reducing energy consumption, energy efficiency improvements make gas and electricity bills more affordable for households and businesses (thereby reducing O&M costs for non-residential customers).[11] Some energy efficiency measures also enable end users to reduce their use of other fuels (e.g. fuel oil, propane, kerosene, biofuel and wood).[12] Together, these factors lead to multiple benefits for energy providers, such as reduced carrying cost on billing arrears; reduced spending on notices and collection agencies; fewer bad debt write-offs; and fewer service disconnections associated with non-payment (Figure 6.4).

These impacts are particularly apparent among low-income customers who often struggle to manage energy bills. Energy efficiency programmes targeted to low-income

10 Key assumptions in the Synapse Energy Economics, Inc. analysis are that the energy providers meet their targets and, in the long run, the price of RECs (and thus the unit cost of RPS compliance) will be determined by the cost of new entry of the marginal renewable energy unit.

11 The Vermont study quantified O&M costs at a NPV of USD 14.8 million, giving a levelised benefit of USD 17.4/MWh.

12 By assessing the avoided fuel supply costs of other fuels, the Vermont study generated a levelised benefit of USD 14.4/MWh.

© OECD/IEA, 2014.

customers have been shown to reduce customer default by 25% or more (Skumatz, 2011). Benefits to the energy provider have been valued at USD 14.7 per household per year; benefits to low-income households were almost double – over USD 27.80 per household per year.

A study conducted as part of a Cincinnati low-income weatherisation programme found similar results. The average arrears of some 2 400 households participating in a community weatherisation programme fell by over 60% following energy efficiency improvements (Drakos, 2013). Evaluation of the US Weatherization Assistance Program (WAP) confirmed this correlation: among participants who responded to the post-intervention survey, the percentage reporting difficulties in paying their bill dropped from almost 75% to 58%, the rate of disconnections was cut in half, and the percentage of those paying less than the full bill amount fell from almost half to just over one-third (Tonn, 2013). In a competitive market, avoidance of image problems associated with disconnecting vulnerable households is also an important benefit, although harder to measure.[13]

Figure 6.4 NEBs for utilities from low-income energy savings programmes

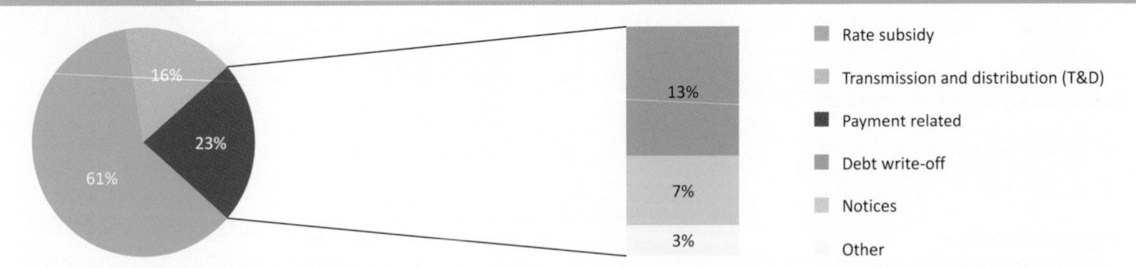

Source: Skumatz (2013), "Non-energy benefits (NEBS): What have we learned in 20 years?", presentation at the IEA Roundtable on Energy Provider and Consumer Benefits, Ottawa, 15-16 October 2013.

Key point *Savings on rate subsidies to low-income customers is the most significant NEB to energy providers; payment-related benefits represent one-quarter of the benefits.*

"Non-energy participant benefits"

Although beyond the focus of this chapter, it is important to note that identifying participant benefits in the assessment of energy efficiency programmes undertaken by energy providers played a key role in raising awareness about the wider multiple benefits such initiatives can deliver. Much of the early quantified evidence of the multiple benefits of energy efficiency first emerged in this context. Energy providers and regulators in Canada and the United States often describe the wider benefits collectively as the NEBs of energy efficiency programmes.

Researchers tend to organise NEBs into three categories: i) those accruing to programme participants (e.g. increased property values, decreased water and sewer bills, increased comfort, health and safety); ii) indirect benefits to the utility (e.g. bill payment improvements, fewer service calls); and iii) societal benefits (e.g. job creation, reduced emissions and health care costs, other environmental benefits). These represent only a subset of all multiple benefits of energy efficiency, several of which are covered elsewhere in this publication.

13 Energy efficiency programmes, particularly those targeting low-income households, can deliver additional benefits such as improved corporate image, enhanced customer loyalty, and closer relations with regulators and government. Energy efficiency activities can also help energy providers establish a price advantage to attract new (or better retain existing) customers.

© OECD/IEA, 2014.

These benefits have been characterised as hard-to-measure (Skumatz, 2006), and many energy regulators consider NEBs to be outside their remit. Although many utility managers and regulators recognise the value of NEBs to be greater than zero, little consensus exists beyond this point. A wealth of case-based evidence indicates that the value of these benefits is extremely high, especially among low-income customers. The California Public Utilities Commission (CPUC) has examined NEBs within the context of analysing the cost-effectiveness of California's demand-side programmes; to date, this method has been used only to evaluate energy efficiency measures that target low-income customers through California's Energy Savings Assistance (ESA) programme, including an initiative by San Diego Gas & Electric (SDG&E)[14] (Table 6.5).

Table 6.5 NEB values for participants of the SDG&E programme

NEB	SDG&E value 2009 (USD)	Value range from SERA study (USD)
Water/sewer savings	7.37	4 to 15
Property value benefits	4.78	3 to 20
Fewer fires	4.64	0.02 to 0.16
Fewer illnesses/lost days from work/school	3.92	4 to 12
Net benefits for comfort and noise	3.56	15 to 20
Net benefits for additional hardship	1.58	
Moving costs/mobility	1.53	< 1
Fewer calls to utility	0.21	0.18 to 0.30
Fewer shutoffs	0.16	0.03 to 12
Fewer reconnects	0.07	0.03 to 0.08
Total	**27.82**	**26 to 80**

Note: Table shows a value for each NEB based on specific characteristics of SDG&E customers and the values found for these benefits from the SERA, Inc. study that analysed a wider range of utility DSM programmes.
Source: Morgenstern, J. (2013), "California's experience in incorporating non-energy benefits into cost-effectiveness tests", IEA Roundtable on Energy Provider and Consumer Benefits, IEA, Natural Resources Canada, Canadian Gas Association and Canadian Electricty Association, Ottawa, 15-16 October, 2013.

Like the CPUC, the Department of Public Utilities in Massachusetts (Mass DPU) has been a frontrunner in the United States, working with programme administrators and evaluators on issues associated with estimating participant benefits (Box 6.7). Mass DPU continues to investigate unresolved issues such as the rigour and transparency of evaluation methods; uncertainty related to interaction of multiple benefits, including the potential for double-counting; consistent application of benefit-cost tests across customers and measures; and associating specific types of benefits with specific measures.

The rapidly growing evidence of multiple benefits for participants in the US context offers an important opportunity for improving the standard practice in evaluation (discussed further in the section on Methodological approaches below).

14 Many examples exist of energy efficiency programmes that target low-income households, including the US Low-Income Home Energy Assistance Program (LIHEAP) and Ireland's *Warmer Homes* scheme (see Chapter 4).

© OECD/IEA, 2014.

Box 6.7 Efforts to measure and evaluate NEBs in Massachusetts

In Massachusetts, Mass DPU first began including multiple benefits for participants (which it calls "non-energy impacts") as part of the programme benefit-cost analysis in 1999. Mass DPU guidelines explicitly include the following NEBs:

- resource benefits that include other fuels saved (e.g. oil, wood, liquid petroleum gas) and water savings
- non-resource benefits, including customer O&M savings (e.g. LEDs in a non-residential environment require less-frequent replacement than original light bulbs), and reduced environmental and safety costs
- all quantifiable benefits for low-income customers.

The recent DPU evaluation indicates that the multiple benefits for energy providers are about USD 16 per low-income programme participant. Over half of this amount is due to reductions from safety-related emergency calls to the energy provider.

Source: Brant, J. (2013), "Including Non-energy Benefits in Evaluating Massachusetts' EE Programs", IEA Roundtable on Energy Provider and Consumer Benefits, IEA, Natural Resources Canada, Canadian Gas Association and Canadian Electricity Association, Ottawa, 15-16 October.

Box 6.8 Demand-reduction induced price effects (DRIPEs)

When investment in energy efficiency reduces overall demand for electricity, it can lead to a drop in the wholesale market-clearing price for electricity.* This derived response is called DRIPE.**

DRIPE estimated values per megawatt hour or megawatt are usually very small, amounting to a fraction of a percent of the annual average market prices of energy. Due to the typically steep supply curve at high loads, the DRIPE per megawatt hour or megawatt impact during peak hours is much higher, which has a further benefit of dampening energy market price volatility. These impacts are projected to dissipate over four to five years as the energy providers react to the new, lower level of energy and capacity required (Hurley et al., 2008).

When expressed in absolute dollar terms (rather than percentages), DRIPE impacts are significant: since the small reductions in market prices apply to all energy being purchased in the market, they can translate into large monetary savings. For energy efficiency measures targeting peak reductions, the associated DRIPE benefits may amount to 15% to 20% of total benefits; this reflects that fact that a small reduction in peak loads may have a larger impact on prices.

In New England, estimates of DRIPE build upon results from a zonal, locational marginal-price-forecasting model that simulates the operation of the energy and operating reserves markets. In addition, the annual New England estimates of the wholesale energy market DRIPE involve an analysis of historical zonal hourly market prices against zonal and regional load. It should be noted that energy DRIPE is applicable only to energy purchased at market prices (Synapse Energy Economics, Inc., 2013). Because there is no additional saving of energy resources from DRIPE, it is classified in this report as a customer benefit.

* Additionally, electric energy efficiency programmes may also reduce natural gas prices for areas where natural gas generation of electricity sets the marginal price for electricity.

** The DRIPE effect is New England nomenclature. In other areas, these effects are described as price mitigation or price suppression.

Benefits to all energy users: wholesale price and other impacts

Improving efficiency of the energy system has the potential to benefit all energy users and society at large through its long-term capacity to lower energy prices. The effect of energy efficiency in reducing the wholesale price of electricity has been well established (Box 6.9);

© OECD/IEA, 2014.

however, some argue that this could be viewed as a transfer payment, in which ratepayers get lower rates while utility shareholders get lower returns.

The benefits from energy efficiency activities will usually be taken into account at future price reviews by the regulator. The reduction in the wholesale price of electricity for all ratepayers can bring energy and financial savings to all consumers within a given market, not just those who participated in the EEOs. In the regulated market case, many of the same benefits will also accrue from improved price setting in the monopoly parts of the provision chain (e.g. passing on to customers the cost savings from delaying forecast investments in the distribution system).

A cumulative view of direct and indirect benefits

The overall benefit to energy providers results from the addition of both direct benefits and those that are generated indirectly through customer benefits. The Efficiency Vermont study illustrates how these benefits can be cumulated (see Box 6.9).

Box 6.9 Efficiency Vermont: Impact of end-user benefits on direct energy provider benefits

As described above, Efficiency Vermont demonstrated that the programme's energy provider multiple benefits were valued at more than USD 104.8/MWh. The additional benefits accruing to participating end users boost the value by at least USD 42/MWh (Table 6.6).

Table 6.6 Multiple benefits accrued to Vermont energy providers from end-user benefits in 2010

Benefit to energy provider	Results (USD/MWh)*
Sub-total of all energy provider multiple benefits	> 104.8
Additional multiple benefits to end users	**Results (USD/MWh)**
Reduced water use	10.8
Saving of other fuels	14.4
Reduced maintenance costs	17.4
Reduced prices in wholesale market	Not evaluated**
Sub-total of additional end-user multiple benefits	> 42.6
Total energy provider and end-use consumer benefits	**> 147.4*****

* For comparison the annual average retail price for electricity in Vermont in 2010 was USD 132/MWh and the average wholesale price in the regional market New England ISO was USD 50/MWh (US EIA, 2011).
** Vermont does not have a binding renewable obligation target expressed in percentage of electricity provided, but the value for other northeastern states has been estimated at USD 1.8/MWh to USD 6.3/MWh.
*** All benefits are levellised in USD 2010 currency rates.
Source: Efficiency Vermont (2012), *Annual Report 2010*, Efficiency Vermont, Burlington, www.efficiencyvermont.com/docs/about_efficiency_vermont/annual_reports/2010_Annual_Report.pdf.

Methodological approaches

The assessment methods described in this section draw on global experience in evaluating impacts within energy provider companies. Many of the examples originate from the United States, where the stringent reporting requirements of a heavily regulated market

© OECD/IEA, 2014.

have necessitated development of a detailed assessment methodology based on cost-effectiveness. Much of the analysis is carried out in the context of EEO systems. Globally, governments have developed similar approaches to evaluating the cost-effectiveness of EEOs. The approaches typically cover three main aspects:

- the full costs to the energy provider of delivering the EEO (subsidies, advertising and marketing expenses, programme recruitment, administering the payment of rebates and reporting, measurement and evaluation, etc.)
- any investment made by end-use consumers, manufacturers, landlords and other third parties
- establishing the net or real energy savings, after allowing for increased amenity, heat replacement effect, estimates of free-riders (those that would have invested in energy efficiency anyway), etc.

Standard cost-effectiveness tests

The *California Standard Practice Manual* establishes standard procedures for cost-effectiveness evaluations for utility-sponsored energy efficiency programmes and is considered an authoritative source for defining cost-effectiveness criteria in utility systems in North America and beyond. The manual sets out five cost-effectiveness tests, three of which are in common usage (as indicated by the percentage of states in which they are applied as the primary test): the Utility Cost Test (71%), the Total Resource Cost (TRC) Test (15%) and the Societal Cost Test (SCT) (12%) (Kushler, Nowak and Witte, 2012).

- The Utility (or Program Administrator) Cost Test examines costs and benefits from the perspective of the utility's revenue requirement. It seeks to answer the question of whether benefits to the utility outweigh the costs of implementation.
- The TRC aims to include all costs and all benefits incurred by participating customers of the utility. It seeks to assess whether the benefits to the utility and participants (together) outweigh the costs. This test includes the full costs of the measure, programme administrative costs and a range of participant benefits (including hard-to-measure benefits).
- The SCT includes all costs and benefits experienced by society as a whole. It seeks to assess whether society is better off with the programme. It includes all of the TRC costs and benefits, plus the benefits of avoiding environmental damage and other externalities. It also uses a lower discount rate to calculate NPVs, reflecting the value of these benefits to future generations.

Challenges with the current cost-effectiveness tests

As noted above, the TRC cost-effectiveness tests cover all programme costs (including participant and third-party contributions) as well as the resultant benefits to the utility, the participating customers, the non-participating customers and the given society as a whole. Individual states do not always apply these tests consistently however, and it is sometimes argued that the TRC is frequently misapplied, as evaluators leave out of calculations many of the costs and benefits to customers (Woolf et al., 2012).

Evaluators and utility managers often face resistance when discussing socio-economic benefits with an energy regulator, due to concerns that the required data are not available and that methodologies are costly to develop and apply. The result has been a classic chicken-and-egg situation: regulators are reluctant to broaden existing frameworks to accept unverified benefit estimates while funding is less available to conduct deeper studies that could investigate and verify multiple benefits (than for studies evaluating the already accepted benefits).

© OECD/IEA, 2014.

A deepened understanding of the range and value of multiple benefits that energy efficiency can deliver for energy providers and their customers highlights the limitations of current frameworks. It also presents an opportunity to further develop standard practice methods to better account for multiple benefits. A fundamental gap is evident in existing evaluation mechanisms and cost-effectiveness tests when it comes to measuring the multiple benefits of energy efficiency for customers. Many experts argue that this has skewed net costs of energy efficiency upwards and lead to systematic under-investment in energy efficiency (NESP, 2014; Synapse Energy Economics, Inc., 2013).

In recent years, evaluators have made significant progress in modifying cost-effectiveness evaluation practices to include benefits other than the avoided costs of demand and energy. In countries that are actively pursuing energy efficiency and related spending is growing quickly, developing more accurate measures of the wider benefits is likely to be a crucial issue for assessing whether to continue such programmes.

In California, some NEBs are now included in the cost-effectiveness tests used for its ESA programme for low-income customers. A model called the Low Income Public Purpose Test was developed, through a collaborative process involving CPUC and the California Public Service Commission, to facilitate estimation of the NEBs per household and determine a programme-wide value of those benefits (CPUC, 2013). The CPUC continues to explore the policy issues associated with introducing NEBs into the standard practice tests, and is considering a new valuation procedure based on the extent to which each ESA measure achieves a particular health or safety improvement (e.g. fewer fires, greater home comfort, improved security).

Box 6.10

A proposal for framework evolution: Resource Value Framework

A group of organisations and individuals working together as the National Efficiency Screening Project (NESP) is developing a new standard practice manual to assist states in improving the methods used to determine the cost-effectiveness of energy efficiency programmes (NHPC, 2014). The goal is to better inform decision makers about which energy efficiency resources are in the public interest and what level of investment is appropriate through the development of improved cost-effectiveness analysis that is consistently applied throughout the United States. The manual presents a *Resource Value Framework* to provide guidance for states to develop and implement tests that are consistent with sound principles and current best practices.

The five main principles include:

- serving the public interest
- achievement of stated energy policy goals
- symmetrical application of tests (both costs and benefits are included in the analysis)
- transparency (through use of standard templates with full information on goals, assumptions and methodologies)
- applicability to all resources.

In general, these principles should be applied to all types of electric and gas utility resources, and should cover both demand- and supply-side resources.

The Resource Value Framework approach recognises that there are essentially two overall regulatory perspectives that utilities may choose when assessing cost-effective energy efficiency: a utility system perspective (that includes all of the costs and benefits experienced by the utility system); and a societal perspective (a more expansive perspective including programme participant costs and benefits). Whichever perspective is chosen, the crucial factor is inclusion of all identifiable relevant costs and benefits. The template recommended for the Resource Value Framework clearly documents the key cost-effectiveness assumptions (e.g. discount rate, measure lifetime, energy savings levels), as well as the quantitative and qualitative benefit and cost findings.

Source: Synapse Energy Economics, Inc. (2014), *Multiple benefits of energy efficiency*, unpublished memo to the IEA.

© OECD/IEA, 2014.

Various players continue to advocate for the inclusion of even broader multiple benefits, such as promoting economic development through energy efficiency activity (see Chapter 2), the value of helping low-income customers pay their energy bills, and realising the energy policy objectives of the particular jurisdiction. A recent and promising development in this area is the attempt by a group of experienced practitioners to develop a modern and comprehensive "standard practice manual" encompassing all the experience gained to date (Box 6.9).

The push for greater monetary quantification of all costs and benefits is already gaining a lot of traction, as is the need for greater transparency on methodology and for alignment of the programme design with the stated energy policy goals. Research in these areas is developing rapidly. The studies and quantified values for multiple benefits of energy efficiency presented in this chapter provide a starting point for further development of existing assessment frameworks. In that context, a careful assessment of actual energy demand reductions generated by EEOs will remain an important starting point, taking into account any rebound effect (Rebound effect perspective 6).

Rebound effect perspective 6

Analysis in the context of energy utilities

When energy efficiency drives down energy demand, the energy unit sales lost by the energy provider can be made up for by other benefits such as capacity cost savings. Energy providers are increasingly moving to absorb any detrimental impact of decreasing unit sales by breaking with the traditional supply-driven approach (selling more electricity or gas units), and moving towards more consumer-focused business models. In this model, the energy provider sells customised energy packages that include – along with the units of energy – energy efficiency-related software, hardware and advice. As interest in energy efficiency grows, energy providers will focus increasingly on delivering energy services rather than only energy unit sales.

Many of the broader range of benefits to the energy provider discussed in this chapter derive from reduced energy demand (i.e. deferral of infrastructure investment, environmental regulation and CO_2 cost savings), and would therefore be lessened proportionate to any rebound effect. This is unlikely to undercut the more enduring systemic benefits represented by the innovative potential of energy service business models, driven by asset valorisation and supporting investor confidence.

Shortcuts: Use of adders

As highlighted in the Companion Guide at the end of this publication of this publication, assessment methods that deliver quantitative results are the most accurate and reliable means of evaluating multiple benefits of energy efficiency. They can, however, require significant time, cost and resources to carry out. In the absence of comprehensive methods for quantifying a broader range of benefits, regulators in Canada and the United States have sometimes adopted "adders" as a proxy to account for the unmeasured participant benefits (such as comfort, health and safety) and societal benefits that have proven difficult to precisely quantify. When adders are quantified as accurately as possible, they can provide simple, reasonable substitutes to quantitative methods for these benefits.

In practice, however, adders are seldom estimated accurately; thus, this approach results in compromises in precision and credibility. There are a range of challenges in using adders, and their applicability is limited due to regional variations in climate, programme types and measures. Adders are therefore set conservatively – typically at 10% to 15% of total measured outcomes – and often underestimate the real value of these benefits (Synapse Energy Economics, Inc., 2014).

© OECD/IEA, 2014.

Recently, as a result of increasing evidence of the multiple benefits of energy efficiency,[15] some regulators are beginning to accept much higher rates for their approved adders. The regulator in British Colombia, Canada, for example, has modified the SCT to include a 30% deemed adder for NEBs in light of the multiplier effect that has been witnessed as a result of broader economic benefits.

In reaction to the imprecise nature of using adders to capture multiple benefits, an alternative method has been used by the regulator in California. Their approach is to make energy efficiency, demand response and renewables "preferred resources" for utilities – and to require utilities to use all available, cost-effective preferred resources before they resort to building or purchasing traditional resources.

Further research for stakeholders

In recent years, the pace and sophistication in widening the range of multiple benefits and attempting to quantify these benefits has been impressive; however, more work remains to be carried out by stakeholders in the energy provider-delivered energy efficiency sphere (Table 6.7).

Table 6.7 Further stakeholder research and collaboration opportunities in energy delivery impacts

Area	Specific actions
Benefit areas and causal linkages	Improve messaging of multiple benefits to match the local circumstances of energy providers and governments to demonstrate to local decision makers the value of improved policy and market design.
	With improved energy security as a major policy issue, undertake more quantified work on the potential role of energy efficiency.
	Conduct more longer-term studies of energy efficiency (like those for New England and Germany) that establish the long-term impact of the range of avoided system costs and social benefits to fully grasp the economic implications of energy efficiency investments.
Data, indicators and metrics	Develop a consistent and robust approach to quantifying the multiple benefits; carry out more innovative evidence-based analysis that is subjected to peer review.
Assessment methodologies	Conduct more longer-term studies of energy efficiency in the energy system to establish better methods for assessing the avoided costs and system outcomes.
Collaboration initiatives	The recently developed Resource Value Framework looks to be a flexible and valuable addition to the methods to identify and quantify multiple benefits from EEOs. In view of the increasing use of EEOs as a policy tool around the world, extending flexible frameworks on a global basis would seem a logical next step. Such a framework could be extended globally with inputs from a number of key agencies and leading energy providers.

Conclusions

Much of the analysis of the multiple benefits aspects for utilities and other energy providers has been motivated by the EEO programmes. Initially, the justification for placing EEOs on energy providers was dominated by comparisons of the value of the energy saving to the resource cost of the energy saved. These calculations were based on a DSM concept that demand-side actions should be undertaken where they are cheaper than the marginal costs

15 In Colorado, the regulator has adopted a 25% adder for low-income programmes and a new proposal is under consideration to increase it to 60%. In 2013, Rhode Island also adopted many of the values from the Massachusetts studies.

© OECD/IEA, 2014.

of new generation. In this context, multiple benefits were always present, but never (or rarely) measured.

Because of the regulatory context of its energy market, the United States has led the way in quantifying multiple benefits and in including the perspectives of the energy provider, the customers who participate in the EEO as well as those who do not, and the societal benefits. Studies show that multiple benefits for the energy provider (and hence ultimately for all customers) are at least as large as the traditional energy resource cost savings. Additional benefits to participating customers are nearly as large again.

An emerging aspect of utility energy efficiency efforts is the changing prospects for demand growth. It is not yet clear what effect declining demand will have on the impacts of DSM programmes. Utilities that pursue energy efficiency may be challenged by their own success in reducing demand, or they may find the pursuit is uneconomical, if energy prices are not dynamic enough to reflect shifting cost structures. As the number of wholesale energy and capacity markets has grown, so has the impact of large-scale energy efficiency and demand response on market prices.

There is growing recognition that the participating customer multiple benefits are particularly important for low-income customers and also benefit the energy provider and all its other customers. Equally important for energy efficiency policy planning is that the multiple benefits to non-participating customers are significant.

The dynamic nature of energy provider multiple benefits highlights the need for standard practice evaluation methods that accurately take into account the full range of costs and benefits that come into play in energy efficiency programmes. Even in a less regulated context, such as that of Europe, better understanding the contribution of energy efficiency to constructing a sustainable energy future will be increasingly important.

Bibliography

Brant, J. (2013), "Including Non-Energy Benefits in Evaluating Massachusetts' EE Programs", presentation at the IEA Roundtable on Energy Provider and Consumer Benefits: IEA study on Capturing the Multiple Benefits of Energy Efficiency, IEA (International Energy Agency), Natural Resources Canada, Canadian Gas Association and Canadian Electricity Association, Ottawa, 15-16 October 2013, www.iea.org/media/workshops/2013/energyproviders/Session3_2_Brant_IEA_Final.pdf (accessed 6 July 2014).

CPUC (California Public Utilities Commission) (2013), *Energy Savings Assistance Program Cost-effectiveness White Paper*, prepared by the ESA (Energy Savings Assistance) Program Cost-effectiveness Working Group, CPUC, San Francisco, http://docs.cpuc.ca.gov/PublishedDocs/Efile/G000/M062/K374/62374022.PDF (accessed 6 July 2014).

CPUC (2001), *California Standard Practice Manual Economic Analysis of Demand-side Programs and Projects*, Governor's Office of Planning and Research, State of California, Sacramento, http://cleanefficientenergy.org/sites/default/files/07-J_CPUC_STANDARD_PRACTICE_MANUAL.pdf (accessed 28 June 2014).

Colburn, K. (2013), "Valuing the environmental benefits of energy efficiency", presentation at the IEA Roundtable on Energy Provider and Consumer Benefits: IEA study on Capturing the Multiple Benefits of Energy Efficiency, IEA, Natural Resources Canada, Canadian Gas Association and Canadian Electricity Association, Ottawa, 15-16 October 2013, www.iea.org/media/workshops/2013/energyproviders/Session25RAP_Colburn_EnvirBenefitsofEE_EIAOttawa_2013_Oct_15.pdf (accessed 6 July 2014).

Craft, R. (2012), "Network deferral benefits of energy efficiency", presentation at the PEPDEE (Policies for Energy Provider Delivery of Energy Efficiency) North American Regional Energy Policy Dialogue, IEA and RAP (Regulatory Assistance Project), Washington DC, 18-19 April 2012, www.iea.org/media/workshops/2013/energyproviders/RebeccaCroftOttawaSection1..pdf (accessed 7 July 2014).

Drakos, J. (2013), "Low-income weatherization benefits for consumers and utilities in Cincinnati, Ohio", presentation at the IEA Roundtable on Energy Provider and Consumer Benefits: IEA study on Capturing the Multiple Benefits of Energy Efficiency, IEA, Natural Resources Canada, Canadian Gas Association and

© OECD/IEA, 2014.

Canadian Electricity Association, Ottawa, 15-16 October 2013, www.iea.org/media/workshops/2013/energyproviders/Session1_5_Drakos_ConsumersandUtilities_PWC.pdf (accessed 7 July 2014).

Eckman, T. (2013), "Realizing the reliability and resource adequacy benefits of energy efficiency in power planning", presentation at the IEA Roundtable on Energy Provider and Consumer Benefits of Energy Efficiency, Ottawa, 15-16 October 2013, www.iea.org/media/workshops/2013/energyproviders/Session2_1_Eckman_IEA_Ottawa_101613.pdf (accessed 7 July 2014).

Efficiency Vermont (2012), *Annual Report 2010*, Efficiency Vermont, Burlington, www.efficiencyvermont.com/docs/about_efficiency_vermont/annual_reports/2010_Annual_Report.pdf (accessed 7 July 2014).

Froehlich, J. et al. (2011), "Disaggregated end-use energy sensing for the smart grid", *Pervasive Computing*, Vol. 10, No. 1, IEEE CS (Institute of Electrical and Electronics Engineers Computing Society), Los Alamitos, www.gabeacohn.com/pdf/Froehlich_DisagEnergy_IEEEpervasive11.pdf (accessed 7 July 2014).

Hurley, D. et al., (2008), *Costs and Benefits of Electric Utility Energy Efficiency in Massachusetts*, report prepared for the Northeast Energy Efficiency Council, Synapse Energy Economics, Inc., Cambridge, www.synapse-energy.com/Downloads/SynapseReport.2008-08.0.MA-Electric-Utility-Energy-Efficiency.08-075.pdf (accessed 7 July 2014).

IEA (International Energy Agency) (2013), *Energy Provider-delivered Energy Efficiency: A Global Stock-taking Based on Case Studies*, Insights Paper, OECD (Organisation for Economic Co-operation and Development)/IEA, Paris, www.iea.org/publications/insights/EnergyProviderDeliveredEnergyEfficiency_WEB.pdf (accessed 7 July 2014).

Kallay, J. (2013), "A new framework for energy efficiency screening", presentation at the IEA Roundtable on Energy Provider and Consumer Benefits, Ottawa, 15-16 October 2013, www.iea.org/media/workshops/2013/energyproviders/Session3_3_Kallay_ANewFrameworkforEEScreening_20131016.pdf (accessed 7 July 2014).

Kushler, M., S. Nowak and P. Witte (2012), *A National Survey of State Policies and Practices for the Evaluation of Ratepayer-funded Energy Efficiency Programs*, research report U122, ACEEE (American Council for an Energy-Efficient Economy), Washington DC, www.aceee.org/research-report/u122 (accessed 3 July 2014).

Lazar, J. and K. Colburn (2013), *Recognizing the Full Value of Energy Efficiency: What's Under the Feel-good Frosting of the World's Most Valuable Layer Cake of Benefits*, presentation at the 2013 ACEEE (American Council for an Energy-Efficient Economy) National Conference on Energy Efficiency as a Resource, ACEEE, Nashville, 22-24 August 2013, RAP (Regulatory Assistant Project), Montpelier, www.raponline.org/document/download/id/6768 (accessed 7 July 2014).

Lazar, J. and X. Baldwin (2011), *Valuing the Contribution of Energy Efficiency to Avoided Marginal Line Losses and Reserve Requirements*, RAP (Regulatory Assistant Project), Montpelier, www.raponline.org/document/download/id/4537 (accessed 7 July 2014).

Morgenstern, J. (2013), "California's experience in incorporating non-energy benefits into cost-effectiveness tests", presentation at the IEA Roundtable on Energy Provider and Consumer Benefits: IEA study on Capturing the Multiple Benefits of Energy Efficiency, IEA, Natural Resources Canada, Canadian Gas Association and Canadian Electricity Association, Ottawa, 15-16 October 2013, www.iea.org/media/workshops/2013/energyproviders/Session3_4_Morgenstern_IEAOct16presentation.pdf (accessed 7 July 2014).

Neme, C. and R. Sedano (2012), *US Experience with Efficiency as a Transmission & Distribution System Resource*, RAP (Regulatory Assistant Project), Montpelier, www.raponline.org/document/download/id/4765 (accessed 7 July 2014).

NESP (National Efficiency Screening Project) (2014), *The Resource Value Framework: Reforming Energy Efficiency Cost-effectiveness Screening*, National Home Performance Council, Home Performance Coalition, Washington DC, www.nhpci.org/publications/NHPC_NESP-Recommendations-Final_20140328.pdf (accessed 7 July 2014).

Opinion Dynamics Corp. (2008), *Indirect Impact Evaluation of the Statewide Energy Efficiency Education and Training Program, Volume I of IV*, final report prepared by Opinion Dynamics Corporation for CPUC Energy Division, CALMAC, San Francisco, www.calmac.org/publications/06-08_Statewide_Education_and_Training_Impact_Eval_Vol_I_FINAL.pdf (accessed 7 July 2014).

Prognos AG, EWI (Energiewirtschaftliches Institut an der Universität zu Köln) and GWS (Gesellschaft für wirtschaftliche Strukturforschung) (2010), *Energieszenarien für ein Energiekonzept der Bundesregierung*,

© OECD/IEA, 2014.

Prognos AG/EWI/GWS, Basel/Köln/Osnabrück, www.ewi.uni-koeln.de/fileadmin/user_upload/Publikationen/Studien/Politik_und_Gesellschaft/2010/EWI_2010-08-30_Energieszenarien-Studie.pdf (accessed 7 July 2014).

Skumatz, L. (2013), "Non-energy benefits (NEBS): What have we learned in 20 years?", presentation at the IEA Roundtable on Energy Provider and Consumer Benefits, Ottawa, 15-16 October 2013, www.iea.org/media/workshops/2013/energyproviders/NEB_Skumatz_OttawaIEA_v6final.pdf (accessed 7 July 2014).

Skumatz, L. (2011), "Co-benefits of Low-Income Weatherization Programs: Framing the Role of Co-Benefits", presentation at the IEA Fuel Poverty Workshop, 27 January 2011, www.iea.org/media/workshops/2011/poverty/pres3_SKUMATZ.pdf (accessed 17 May 2014).

Skumatz, L. (2006), *Measuring "Hard to Measure" Non-energy Benefits (NEBs) from Energy Programs: Methods and Results,* American Evaluation Association, Portland.

Skumatz, L. (1997), "Recognizing All Program Benefits: Estimating the Non-Energy Benefits of PG&E's Venture Partner's Pilot Program (VPP)", *Proceedings of the 1997 Energy Evaluation Conference,* Chicago, Illinois, 1997.

Smith, S. (2013), "Totally Radically Cool: A new approach to DSM", presentation at the IEA Roundtable on Energy Provider and Consumer Benefits of Energy Efficiency, Ottawa, 15-16 October 2013, Fortis BC Energy Utilities, http://www.iea.org/media/workshops/2013/energyproviders/Session3_5_Smith_EPMBOttawa.pdf (accessed 21 June 2014).

Synapse Energy Economics Inc. (2014), *Multiple Benefits of Energy Efficiency*, unpublished memo to the IEA.

Synapse Energy Economics Inc. (2013), *Avoided Energy Supply Costs in New England: 2013 Report*, report prepared for the Avoided-Energy-Supply-Component (AESC) Study Group, Synapse Energy Economics, Inc., Cambridge, www.synapse-energy.com/Downloads/SynapseReport.2013-07.AESC.AESC-2013.13-029-Report.pdf (accessed 7 July 2014).

Tonn, B. (2013), "Making sense of non-energy benefits: Results from the Weatherization Assistance Program", presentation at the IEA Roundtable on Energy Provider and Consumer Benefits: IEA study on Capturing the Multiple Benefits of Energy Efficiency, IEA, Natural Resources Canada, Canadian Gas Association and Canadian Electricity Association, Ottawa, 15-16 October 2013, www.iea.org/media/workshops/2013/energyproviders/Session1_4_Tonn_CanadaIEANEBs.pdf (accessed 7 July 2014).

Tyrell, C. (2013), "Integrating Conservation and Demand Management into Distribution Operations", presentation at the IEA Roundtable on Energy Provider and Consumer Benefits of Energy Efficiency, Ottawa, 15-16 October 2013, Toronto Hydro, http://www.iea.org/media/workshops/2013/energyproviders/Session1_2_Tyrell_CDMIntegrationIntoDxOpsOct152013FINAL.pdf (accessed 21 June 2014).

UK DECC (United Kingdom Department of Energy & Climate Change) (2014), *The Future of the Energy Company Obligation: Consultation Document*, United Kingdom government, London, www.gov.uk/government/uploads/system/uploads/attachment_data/file/291900/Energy_Company_Obligation__ECO__The_Future_of_the_Energy_Company_Obligation_Consultation_DocumentFINAL.pdf (accessed 6 July 2014).

UK DECC (2008), *Evaluation of the Energy Efficiency Commitment 2005-08*, final report prepared by Eoin Lees Energy, United Kingdom government, London, http://s3.amazonaws.com/zanran_storage/www.defra.gov.uk/ContentPages/4234041.pdf (accessed 7 July 2014).

US IEA (United States Energy Information Administration) (2011), *Electric Power Monthly – January 2011*, US Department of Energy, Washington DC.

Wisniewski, E. (2013), "Developing a Binational Framework for Connected Homes", presentation at the IEA Roundtable on Energy Provider and Consumer Benefits, Ottawa, 15-16 October 2013, www.iea.org/media/workshops/2013/energyproviders/Session2_2_Wisniewski_Canada_ConnectedHome.pdf (accessed 7 July 2014).

Woolf, T. et al. (2012), *Energy Efficiency Cost-Effectiveness Screening: How to Properly Account for "Other Program Impacts" and Environmental Compliance Costs*, report prepared by Synapse Energy Economics, Inc. for RAP and the Vermont Housing Conservation Board, RAP, Montpelier, Vermont, www.raponline.org/document/download/id/6149 (accessed 7 July 2014).

Wünsch, M. et al. (2014), *Benefits of Energy Efficiency on the German Power Sector*, Agora Energiewende, Berlin, www.agora-energiewende.org/fileadmin/downloads/publikationen/Studien/Energieeffizienz/Agora_ECF_RAP_System_Benefit_Study_short_version_web.pdf (accessed 23 June 2014).

© OECD/IEA, 2014.

Chapter 7

Conclusion: A new perspective on energy efficiency

Key points

- *Tailoring the multiple benefits approach to individual country contexts will maximise its capacity to support national or sub-national priorities for economic and social development. When adopting this approach, governments should take into account that different stakeholders will value the various benefits differently.*
- *Wider implementation of energy efficiency requires capacity building among a broad range of actors, both within the policy domain and in technical and service provider spheres. To maximise the prioritised benefits of a particular energy efficiency policy or programme, greater effort is needed to effectively communicate to diverse actors and audiences, including the general public.*
- *A strength of the multiple benefits approach is that it encourages cross-sectoral collaboration in policy making in which energy efficiency can play a central or supporting role. By optimising the intellectual, technical and financial resources available, it allows governments to tackle more complex issues in a more holistic manner.*
- *While active investigation of the multiple benefits of energy efficiency is still in its early stages, the significant values identified to date demonstrate enormous potential. Concerted efforts to continue building the evidence base could initiate a step-change in the uptake of energy efficiency opportunities, and reposition energy efficiency as a mainstream policy tool for economic and social development.*

Introduction

The analysis and case studies presented in the preceding chapters represent an important expansion and improvement of standard energy efficiency policy process – and the culmination of efforts over several decades to understand better the full range of impacts and benefits that energy efficiency can generate. Collectively, they contribute to strengthening the evidential foundations for bringing the multiple benefits approach more fully into the mainstream policy process. With better information and clear demonstration of the potential to identify, measure and quantify both tangible and intangible impacts arising from energy efficiency measures, policy makers are in a better position to make the case for including these impacts in decision making for energy efficiency policy.

If applied widely, this approach can deliver substantial economic and social value across the five areas covered, and in sectors beyond the scope of this publication. While recognising that much of the application of the multiple benefits approach will be sector-specific, the

© OECD/IEA, 2014.

consultative process through which this publication was developed also brought to the fore several overarching thematic strands. Ultimately, energy efficiency policies need to be tailored to the country context, and support the national or sub-national priorities for economic and social development. A key strength of the multiple benefits approach is that it encourages a cross-sectoral approach to policy making, and can enhance the capacity of governments to tackle more complex issues through interdisciplinary co-operation.

As it is a relatively new field, governments need to consider related issues, such as building capacity for implementation of energy efficiency measures and how to communicate to all stakeholders – including the general public – that a novel approach may have inherent uncertainties, but shows strong potential to delivery greater value. Some of the challenges and opportunities presented by the multiple benefits approach are outlined below.

Optimising the multiple benefits approach

During the course of exploratory studies some lessons have been learned about how best to put a more holistic, outcome-based approach into practice, harnessing new opportunities for gathering, applying and communicating information about the value of energy efficiency. Some early insights which are likely to be valuable in applying a multiple benefits approach and managing the opportunities and challenges it presents are offered below.

Different country perspectives on benefits

National circumstances, economic and social priorities, and the existing system within which a policy operates will each play important roles in the uptake of the multiple benefits approach and the outcomes it can deliver. It is likely to play out differently in every country. Policy makers will need to assess the type and scope of possible multiple benefits that could realistically be achieved in their own countries, taking into account determining factors such as geographic situation, level of economic development, energy resource endowments and demographics. Countries in which efficiency in buildings is already high, for example, may not expect a major improvement in health through building energy efficiency measures; rather, they might focus on the potential to support employment through investment in energy efficiency.

Similarly, local needs and challenges (in light of the socio-economic context and the policy backdrop of the country) will drive the outcomes that policy makers seek to target and maximise through policy design. Some countries may choose to prioritise energy affordability and maximise the role of energy efficiency measures in reducing costs for energy consumers. Others may focus on the potential to boost productivity in certain industrial sectors. The choice of priority benefits should be guided by the strategic or annual priorities of the government to avoid any detrimental overburdening of individual policies or conflict or confusion among priorities. It is equally true that national priorities will determine whether a specific benefit is considered a primary aim or a co-benefit, and thus the value attributed to it by different stakeholders (Table 7.1).

Existing administrative and institutional frameworks will also influence how governments choose to integrate multiple benefits considerations. Political mandates, previous political choices or political expectations of domestic institutions and stakeholders will play a role, as will any constraints created by the domestic legal and institutional context.

The appropriate methodology for policy assessment will also be affected by the local context. Similarly, the degree to which certain multiple benefits are included in impact assessment approaches might, in practice, be dictated by the availability of relevant data in the country. Ideally, identification of data gaps will lead to expansion of existing

© OECD/IEA, 2014.

data collection efforts and to adaptation of administrative systems to support a holistic integration of the multiple benefits approach into the policy process. As resources are not always readily available to support exploratory efforts, opportunities to adapt existing mechanisms and draw appropriate indicators out of existing data sets could offer valuable starting points. Initial efforts made using existing data can be used to support bids to further develop approaches and seek new data sources. Once administrative and data requirements have been identified, additional capacity can be built up over time.

Early identification of the desired outcome objectives will enable policy makers to quickly draw on best-practice examples (i.e. those contained in this report and elsewhere), and supplement these with local knowledge to design energy efficiency policy that maximises the targeted benefits.

Table 7.1 Illustrative valuation of benefits in relation to varying priorities of stakeholders

Benefits vs. co-benefits → multiple benefits				
	Country or stakeholder A	Country or stakeholder B	Country or stakeholder C	Etc.
Industrial competitiveness	Co-benefit			
Fuel imports	Primary	Co-benefit		
Poverty alleviation and development			Primary	
Reducing greenhouse gas (GHG) emissions		Primary	Co-benefit	
Employment creation	Co-benefit	Co-benefit		
Local pollution	Primary		Co-benefit	

Source: Unless otherwise noted, all material in figures and tables in this chapter derives from International Energy Agency (IEA) data and analysis.

Key point *Individual countries and stakeholders are likely to view individual benefits as being of different value.*

Interdisciplinary co-operation

Many governments currently operate in silos, with expert teams working on their mandated policy issues in an isolated way. In such contexts, decision making on energy efficiency has typically been entrusted entirely to energy experts. As governments become more aware of energy efficiency's capacity to deliver outcomes across a range of policy areas, a compelling case arises for creating policy-specific teams that bring together the skills and experience of experts from diverse fields so decision makers have full and accurate information on which to base their decisions.

Policies targeting broad objectives can be devised collaboratively with the engagement of the various ministries having a stake in the outcomes. Such cross-disciplinary co-operation can enable the introduction of new skill sets and alternative perspectives to contribute relevant insight about potential impacts. In energy efficiency, a multiple benefits approach implies the involvement of economists, health experts, business strategists and social services.

© OECD/IEA, 2014.

Many examples exist in which systematic communication among ministries is already underway. This approach has been used to support linking GHG emission reduction goals with energy efficiency policy. Finland's Ministerial Working Group on Climate and Energy Policy, for example, is made up of representatives from various other ministries including foreign affairs, finance, trade and industry, agriculture and forestry and transport and communications. This enables collaborative planning of sectoral policies to ensure that various initiatives are coherent and complementary (IEA, 2013).

While governments routinely call in scientific and technical expertise to help with certain policy questions, in a multiple benefits approach such consultation becomes a more systematic part of policy processes, from planning to evaluation phases. The European Commission has produced guidelines for collecting and using external advice. Interesting examples can be found of specialised community-level agencies playing a role in delivering energy efficiency policies, for example engaging social workers and medical services to identify families who would benefit from energy efficiency interventions. In France, a novel energy efficiency driven initiative established a role for a Medical Indoor Environment Counsellor[1] who is trained to investigate cases in which medical doctors suspect that health symptoms witnessed in a patient are linked to his/her indoor environment (De Blay et al., 2003).

Drawing on expertise from a range of disciplines will require a clear governance structure to ensure that multiple inputs are managed effectively. The European Commission guidance notes that "consultation is not a one-off event, but a dynamic process that may need several steps" (EC, 2009) and while many players could be involved, one ministry might take central responsibility for managing inputs from different contributing bodies and taking the process through to completion. This supports coherency of government activities and facilitates knowledge and data sharing across government, tapping into best-practice approaches wherever they lie. The same is true when implementing energy efficiency policies (Box 7.1).

Box 7.1 "Zip-Up" implementation method

In large-scale programmes, it is particularly important to develop co-ordinated delivery plans that ensure multiple objectives are met in a systematic way. The *Kirklees Warm Zone* project was a major regional energy retrofit scheme administered over three years with the objective of reducing fuel poverty in the borough of Kirklees, England. It was a ground-breaking project, not only due to its novel approach to delivery of energy efficiency measures across a large implementation area, but because of the concrete results it delivered about mental health impacts of energy efficiency.

The value of using of a range of technical and community resources was exemplified in the Zip-Up approach that was used in delivery of the Kirklees programme. Drawing on a range of government, commercial and community bodies, implementation was designed to ensure that the right houses were reached with the right measures in an appropriate order (taking into account questions of social equity). Retrofits were implemented on ward-by-ward basis, while giving priority treatment to vulnerable houses identified at the outset. Ultimately, this approach delivered 30% to 50% higher efficiency than a conventional approach.

Source: Liddell, Morris and Langdon (2011), *Kirklees Warm Zone. The Project and Its Impacts on Wellbeing*, report commissioned by the Department For Social Development Northern Ireland, University of Ulster, Coleraine.

1 www.cmei-france.fr/index.php?section=1-accueil-du-site-des-cmei.

© OECD/IEA, 2014.

Co-operative approaches can deliver higher returns on investment both in savings and improved outcomes for citizens; they also leverage potential economies of scale. When an evaluation commissioned by the New Zealand Ministry of Economic Development (see Chapter 4) returned results about the massive health and well-being impacts of energy efficiency retrofits in low-income homes, other ministries, including the Ministry of Health and Treasury were highly responsive (Grimes et al., 2011). The results supported a budget allocation under the annual Treasury funding round, for a new three-year energy efficiency programme specifically targeting low-income households, particularly families with children and high health needs. In planning the programme, the Ministry for Economic Development (responsible for energy efficiency) consulted with Ministry of Business, Innovation and Employment, the Ministry of Health and Treasury. The programme is to be delivered as a series of projects in partnership with industry, health agencies, specialist community organisations and iwi[2].

Designing and implementing truly integrated policies across different government departments is not without challenges and the potential need for additional co-ordination effort should be taken into account. Nevertheless, it should deliver a deeper understanding of the full range of dynamics at work when energy efficiency improves, and improve policy design to maximise desired outcomes.

Capacity building

More robust projection and subsequent evaluation of the impacts of energy efficiency necessitates, in most countries, additional education and training of policy professionals – but these capacity-building efforts should also target professionals involved in delivering, installing and monitoring energy efficiency measures. Relevant stakeholders need to be made aware of the various dynamics that can be affected by energy efficiency improvements, and of the need to take a holistic approach to account for the interdependence of various energy efficiency measures. All contributors need to understand how to execute their roles to maximise desired impacts and minimise negative ones (for example, alerting installers to the risk that poorly implemented retrofit measures could impair indoor air quality).

Actors that could play a role in a multiple benefits approach to energy efficiency include:

- technicians
- energy service providers
- commercial vendors of energy-related products and services
- health professionals
- real estate agents, building owners and developers
- social workers
- energy auditors
- financial lenders.

Some of these professionals and technical experts represent an important interface with the beneficiaries of energy efficiency measures and play important roles in sensitising the public to the multiple benefits and costs at stake.

2 *Iwi* is a Maori word meaning *tribes*, and refers to the various Maori tribes of New Zealand who are consulted, in accordance with statutory requirements, in the context of many New Zealand government proposals.

© OECD/IEA, 2014.

Box 7.2 Getting the energy efficiency message across to households

In the context of its residential renovation and retrofit programme, the Renovate Right campaign in Canada made a concerted effort to speak directly to householder-parents about what matters most to them. It effectively delivered messages regarding the need to ensure that home repair or energy upgrades are implemented in a way that maximises the positive impacts on child health and ensures the avoidance of harm. The campaign targeted tenants (renters) in particular, with a special focus on low-income households, given the nation-wide correlation between low-income and older housing (and the risks associated with older housing including lead, asbestos and other contaminants).

The campaign included interactive online resources and leaflets that were translated into seven of the most commonly spoken languages in Canada (Cooper, 2013). The team even found an audience on renovation shows on television, and achieved major public dissemination of the healthy retrofit message.

Project managers delivering Renovate Right renovations or retrofits took a collaborative approach, engaging with municipal building departments, utilities delivering energy efficiency retrofits and retailers to ensure the coherence and consistency of the initiative.

Source: Canadian Environmental Law Association (CELA) (2011), *Healthy Retrofits : The Case for Better Integration of Children's Health Protection into Energy Efficiency Programs*, CELA, Toronto.

Communication and messaging

The multiple benefits approach to energy efficiency policy provides a new context for talking about energy efficiency, one which will have relevance to a much wider audience. Traditionally, energy efficiency has been a straight calculation of units of energy saved or energy costs avoided, of interest only to energy experts. This discourse has expanded to other specific areas, such as job creation or GHG emissions reduction, but this expansion has not been systematic or consistent. With the multiple benefits approach, energy efficiency can be transformed into a story of social and economic development that supports a future that is at once prosperous and sustainable. In developed countries, this could help to disassociate energy efficiency from its common perception as a stoic choice to deprive oneself of life's pleasures in order to serve the greater good, a misunderstanding that has arisen from many past messages about the urgent need to restrict energy use to a minimum. In developing countries, it can be more appropriately positioned as a pro-growth instrument, not one designed to limit the development of their economic and indigenous resource base.

As reflected in the multiple benefits approach, different audiences will be interested in different aspects and present distinct communication opportunities. Consistent with the range of potentially interested audiences, there is an opportunity to tailor policy messages to respond to their varying interests (Table 7.2).

Communicating the multiple benefits of energy efficiency to non-experts is particularly important. Some powerful tools, such as indexes and labels, have already been developed. In the United Kingdom, for example, the *Housing Health and Safety Rating System* (*HHSRS*)[3] has allowed professionals from both housing and health disciplines to pool expertise and raise the profile of this issue. The value of the HHSRS in England has created interest in replicating it elsewhere: the United States and the European Union are all undertaking similar efforts. Another idea is to redevelop existing energy labelling to include basic

3 United Kingdom Government, Department for Communities and Local Government (DCLG): London (2006), *Housing Health and Safety Rating System: Guidance for Landlords and Property Related Professionals*, DCLG Publications, Wetherby. www.gov.uk/government/publications/housing-health-and-safety-rating-system-guidance-for-landlords-and-property-related-professionals.

© OECD/IEA, 2014.

parameters about how the product (or service) contributes to multiple benefits. In addition to increasing awareness, such instruments could support possible premiums on more energy efficient assets.

Use of novel but relevant comparators can help to convey messages about multiple benefits in a political context. For example, comparing the costs of energy efficiency upgrades with the cost of "one night in hospital" for health impacts was trialled in the context of the *Warm Up New Zealand: Heat Smart* programme. This implicitly asked people and, more importantly, governments, to decide which they prefer to invest in. Framing "warmth as medicine" is another powerful message about the impact an energy efficient home can have on health (Howden-Chapman, 2013).

Table 7.2 Potential areas of interest among stakeholders in the multiple benefits approach

Type of stakeholder	Likely areas of interest in understanding the wider impacts of energy efficiency
General public	Improved quality of life overall. Reduced energy costs and increased disposable income. Improved health, well-being, productivity, economic status, etc.
Companies and businesses	Improved profitability. Increased competitiveness. Reduced risk profile. Greater sustainability of the service provided to customers. Enhanced "green" image. Better understanding of value of capital investment in energy efficiency. Enhanced business models and/or business case for energy efficiency projects.
Energy efficiency programme or policy level	Enhanced ability to design energy efficiency policy instruments that meet other policy objectives (beyond energy savings). Enhanced ability to provide more comprehensive assessment of results and impacts from programmes or policies. More comprehensive public accountability on policy outcomes. New knowledge that can be used to reshape public policy. Improved justification for resources for new programmes or for programme continuation or expansion.
Policy planning and strategy	Justification for investing in policies to promote energy efficiency. Improved basis for decision-making on where to allocate resources (e.g. energy efficiency, new generation or other measures). Awareness of role of energy efficiency in delivering other outcomes. Deeper understanding of economic growth drivers.

Key point *Different stakeholders will be interested in different aspects of multiple benefits and present distinct communication opportunities.*

Deepening the understanding of multiple benefits

Evidence gathered so far is testimony to the value of an enquiry into multiple benefits. But the active investigation of these outcomes is still in its early stages. The most compelling message that emerges from this investigation is that much more work needs to be done to fully understand the interactions that occur across the economy and society through investments in energy efficiency.

© OECD/IEA, 2014.

Evaluators have started to understand these dynamics, yet have also confirmed the need to dig deeper. In addition to the next steps identified within each of chapters 2-6, several high-level topics emerge as meriting further investigation in a cross-cutting way. These tend to fall across six broad categories of effort:

- strengthen research efforts in general
- enhance data and information collection initiatives
- develop information-sharing mechanisms
- develop methodologies for assessment, quantification and monetisation of impacts
- establish guidelines and tools to ensure information is comparable and transferable
- build capacity for all stakeholders in the policy process.

Conclusion

The valuation of multiple benefits is "a challenging but essential element of policy assessment that should be attempted wherever feasible" (UK HM Treasury, 2003). Energy efficiency policy experts, especially evaluators, are alert to the multiple benefits of energy efficiency, but reliable tools to explore them in depth have been lacking – as has the mandate to do so. Nevertheless, the early results presented in preceding chapters offer the opportunity and indicate growing interest to push the multiple benefits of energy efficiency up the political agenda.

Internationally, this is evidenced by the number of new multiple benefits studies recently initiated, including that commissioned by the Directorate-General for Energy in the European Commission. Studies showcased at recent editions of the International Energy Policy Evaluation Conference and a range of investigations being carried out at the national level are also encouraging.

A multiple benefits approach acknowledges the role of energy efficiency as a fundamental enabler of economic and social development. Fuller consideration of the various impacts, positive but also negative, will assist policy makers in deciding how to allocate resources (whether financial, institutional, administrative or political will) across different policy areas.

This endeavour requires innovation in several areas: the way governments design policies; the degree to which stakeholders are engaged in both designing and implementing policies; and the technical and political standards by which policy success is measured. It calls for greater attention to policy priority setting, and especially to careful monitoring and evaluation of the outcomes of energy efficiency policies. Increased efforts to refine existing metrics and tools (and to develop new ones) will support increasing consistency of approaches internationally. This will, in turn, facilitate data sharing across projects and countries, and enable larger meta-analyses to be carried out, further building confidence in the multiple benefits approach and its results.

This publication reflects concerted effort by many analysts to learn more about the diverse outcomes of energy efficiency policy and, specifically, to find ways to capture the value of the multiple benefits it delivers. The emerging body of knowledge is vital to short-term arguments to adopt a multiple benefits approach using existing analytical practices. Yet the IEA also seeks to encourage future expansion of this new frontier. The results in this publication are not presented as final, but rather designed to deepen the discussion among analysts and other stakeholders in energy and other disciplines to find better ways to identify and measure impacts and to objectively evaluate energy efficiency policy outcomes.

© OECD/IEA, 2014.

Ultimately, this could enable the multiple benefits approach to be integrated systematically into energy efficiency policy assessments and boost understanding among policy makers and the public about the real value of improving energy efficiency. Applying a multiple benefits approach to energy efficiency policy enables a fuller understanding of the potential of energy efficiency to generate tangible improvements in economic and social development. It signals a shift away from the traditional view of energy efficiency as simply delivering invisible energy demand reductions, and recognises its important role within the portfolio of mainstream economic policies for economic and social development. The economic and social signals a multiple benefits approach creates could help to shift energy efficiency from its status as the "hidden fuel" to a recognition of its role as the "first fuel".

Bibliography

Cooper, K. (2013), "The 'Healthy Retrofits' Project: Avoiding children's health risks during energy retrofits and home renovations in Ontario, Canada", presentation at the IEA-EEA Roundtable on the Health & Well-being Impacts of Energy Efficiency Improvements, Copenhagen, 18-19 April 2013, www.iea.org/workshop/roundtableonthehealthwell-beingimpactsofenergyefficiencyimprovements.html (accessed 22 June 2014).

De Blay, F. et al. (2003), "Medical Indoor Environment Counsellor (MIEC): role in compliance with advice on mite allergen avoidance and on mite allergen exposure", *Allergy*, Vol. 58, No. 1, Wiley-Blackwell, Copenhagen, pp. 27-33, http://onlinelibrary.wiley.com/doi/10.1034/j.1398-9995.2003.23674.x/abstract (accessed 28 June 2014).

EC (European Commission) (2009), *Impact Assessment Guidelines*, Sec(2009) 92, EC, Brussels, http://ec.europa.eu/smart-regulation/impact/commission_guidelines/docs/iag_2009_en.pdf (accessed 22 June 2014).

Grimes A. et al. (2011), *Cost Benefit Analysis of the Warm Up New Zealand: Heat Smart Program*, report for the Ministry of Economic Development, Motu Economic and Public Policy Research, Wellington, www.healthyhousing.org.nz/wp-content/uploads/2012/05/NZIF_CBA_report-Final-Revised-0612.pdf (revised June 2012, accessed 22 June 2014).

Howden-Chapman, P. (2013), "Capturing the health benefits of energy efficiency programs in New Zealand – a success story", presentation at the IEA-EEA Roundtable on the Health & Well-being Impacts of Energy Efficiency Improvements, Copenhagen, 18-19 April 2013, www.iea.org/workshop/roundtableonthehealthwell-beingimpactsofenergyefficiencyimprovements.html (accessed 22 June 2014).

IEA (International Energy Agency) (2013), *Energy Policies of IEA Countries: Finland*, OECD (Organisation for Economic Co-operation and Development)/IEA, Paris, www.iea.org/w/bookshop/add.aspx?id=452 (accessed 28 June 2014).

Liddell C., C. Morris and S. Langdon (2011), *Kirklees Warm Zone. The Project and Its Impacts on Well-being*, report commissioned by the Department For Social Development Northern Ireland, University of Ulster, Coleraine, www.kirklees.gov.uk/community/environment/energyconservation/warmzone/ulsterreport.pdf (accessed 28 June 2014).

NZ Treasury (Treasury, New Zealand Government) (2010), *Cross Agency Initiatives Process Guide for Public Service Agencies*, government of New Zealand, Wellington, www.treasury.govt.nz/publications/guidance/planning/caip (accessed 16 March 2014).

Patterson, C. (2013), "Capturing health benefits for New Zealand's vulnerable populations", presentation at the IEA-EEA Roundtable on the Health & Well-being Impacts of Energy Efficiency Improvements, Copenhagen, 18-19 April 2013, www.iea.org/workshop/roundtableonthehealthwell-beingimpactsofenergyefficiencyimprovements.html (accessed 22 June 2014).

Tetra Tech, Inc. and MPA (Massachusetts Program Administrators) (2012), *Final Report – Commercial and Industrial Non-energy Impacts Study*, June 20, 2012, Tetra Tech, Inc., Madison, http://www.rieermc.ri.gov/documents/evaluationstudies/2012/KEMA_2012_MA_CI_NEI_REPORT.pdf (accessed 17 May 2014).

UK HM Treasury (United Kingdom, Her Majesty's Treasury) (2003), *The Green Book: Appraisal and Evaluation in Central Government*, United Kingdom government, London, https://www.gov.uk/government/uploads/system/uploads/attachment_data/file/220541/green_book_complete.pdf (accessed 28 June 2014).

© OECD/IEA, 2014.

Companion guide

Methodologies for the multiple benefits approach

Key points

- *Identifying, quantifying and assessing the multiple benefits of energy efficiency may seem daunting; in fact, as case studies in this publication demonstrate, much of the work can be done with existing methods, tools and modelling platforms. The complexity inherent in multiple benefits may require using familiar tools in combination, adapting them slightly, or learning to integrate proven methodologies from other disciplines.*
- *Monetising the outcomes of energy efficiency policies is particularly challenging in that relatively few benefits are both objective and quantitative. While valuing benefits that are more subjective and qualitative is inherently less robust, the available methods enable evaluators to prove an important point: even a rough estimate of a benefit value is more accurate than assuming a value of zero.*
- *Once estimated, values for multiple benefit impacts can be integrated into traditional policy decision-making tools such as benefit-cost analysis (BCA), calculation of payback periods and multi-criteria decision analysis (MCDA). Different tools make it possible to integrate the multiple benefits approach at each stage of policy planning, implementation, monitoring and evaluation.*

The multiple benefits approach introduces new challenges and opportunities across the entire process of energy efficiency policy development, from planning and implementation to monitoring and evaluation. Evaluating the impacts of energy efficiency policies is the most essential element, and often the one which policy makers spend the least time on.

Many of the assessment tools and methods commonly used for assessing traditional energy efficiency benefits – energy demand reduction and lower greenhouse gas (GHG) emissions – can be applied to the multiple benefits approach. Early work in this area shows that some tools and methods are better suited than others. Ongoing efforts to put them into practice in the multiple benefits context will support development of increasingly tailored methods for measuring multiple benefits – and indeed may lead to innovation in assessment methods.

Most governmental policy assessment guidelines recommend consideration of "non-market impacts" that cover a range of social, economic and environmental issues, and several governments have published authoritative reference materials. For example, in the United Kingdom, the *Green Book* offers a benchmark for best practise in managing and monitoring

© OECD/IEA, 2014.

policy impacts throughout the policy cycle and emphasises the challenging but essential task of assessing multiple benefits (UK HM Treasury, 2003). The *Impact Assessment Guidelines*, produced by the European Commission, focus largely on considering policy options prior to implementation, and recommend analysis methods that can integrate a mixture of qualitative, quantitative and monetary data, with varying degrees of certainty (EC, 2009). The *California Standard Practice Manual*, which is used throughout the United States, presents five cost-effectiveness tests, including a Societal Cost Test (SCT) that integrates societal discount rates and externality costs or "non-energy co-benefits", also known as "adders" (CPUC, 2001).

To date, multiple benefits assessment remains limited for two reasons: methods for assessing the costs and benefits of non-market impacts have not been fully developed; and such assessments require greater resources (financial and human) than more traditional policy assessment.

This guide describes various policy assessment tools and demonstrates different ways to attribute value to multiple benefits, either individually or cumulatively, in a context of varying levels of information and experience. Brief case studies from macroeconomics, public budgets, health and well-being, industry, and energy delivery benefits are presented to illustrate how the impact of multiple benefits can be integrated into energy efficiency policy decision making. Better understanding of decision-making tools shows how a multiple benefits approach complements existing analysis. Enhancing evaluation within a multiple benefits approach can assist in two ways, by:

- building the evidence base to measure more diverse outcomes in quantitative, monitised or qualitative terms
- facilitating a more holistic approach to policy decision making.

The existing context for energy efficiency policy assessment

Two points in the policy process are of particular relevance to capturing the multiple benefits of energy efficiency: *ex ante* appraisal of policy options (assessing and choosing between various policies) and *ex post* policy evaluation (measuring the impacts of a chosen policy). In either case, measuring the net energy demand reduction generated as a result of the policy remains a useful starting point. This represents one potential benefit in itself, and additional benefits which are generated by energy demand reduction can then be calculated accordingly. Some benefits will be calculated independently of energy savings.

Defining a baseline scenario is a critical first step in any effort to assess the impact of an energy efficiency policy. The baseline is an estimate of the energy that would be needed in the absence of energy efficiency policy: as such, it "sets the bar" against which actual energy demand can be compared to provide a means of measuring impacts of the policy intervention. Different methods can be used to set the baseline, the most common being straightforward examination of the energy consumed before implementation of an energy efficiency policy. The choice of which method to use tends to vary depending on the sector and the measure being assessed, as well as practical constraints on time and resources.

Ultimately, the assessment compares the estimated levels of energy demand against any reduction resulting from the policy, assuming the change reflects energy-user responses.

Energy efficiency assessments should, at the outset, establish a timeframe over which the savings will be monitored and build in a method for accounting "accrued" savings to reflect

© OECD/IEA, 2014.

that an energy efficiency measure will deliver an annual quantity of energy cost saving, as well as cumulative savings over the lifetime of the efficient good or service.

Traditional assessments comprise a simple calculation of gross energy cost saving delivered by an energy efficiency measure, which largely focus on measuring changes in energy demand, units of energy saved and GHG emissions reduced. All of these outcomes can be quantified and linked to market values with relative ease. Increasingly, experts argue assessments should focus on *net* energy demand reduction – i.e. counting only savings attributable to the measure. This approach also makes it possible to account for impacts of free-ridership, spill-over and induced market effects[1] (Kushler, Nowak and Witte, 2012).

This traditional approach typically translates the resulting energy demand reduction into a monetary value, which requires assumptions about issues such as conversion losses, fuel mix and energy prices. Monetisation of the results helps build a strong case for energy efficiency decision making in both public and private sector contexts. A key challenge in multiple benefits is to find mechanisms to monetise less tangible outcomes, or indeed to define different value systems when assigning a dollar figure simply is not practical.

A robust policy assessment should take into account the full range of challenges and confounding factors. In the multiple benefits approach, estimating (even imperfectly) the potential rebound effect(s) at an early stage is vitally important – particularly when energy demand reduction or GHG emission reduction goals are linked to the policy. Including rebound factors[2] in forecasts of energy demand reduction from energy efficiency begins to capture the rebound effect. However, the rebound effect is influenced by the time, space, policy, economy and sector in which it arises (Turner, 2013), making its assessment complex. While various attempts have been made to assess rebound effects, to date there are few examples in which the rebound effects of energy efficiency have been fully analysed and estimates vary widely. Continued effort to obtain good data on the rebound effect will be important for policy makers and evaluators, in order to design policy that maximises energy demand reduction and positive rebound effects while minimising undesirable outcomes (see the discussion of the rebound effect in Chapter 1).

An assessment of energy demand reduction should clearly identify any unresolved issues (particularly the various effects described above) so they can be managed in the energy efficiency policy process. While methods vary, there is a strong body of knowledge on estimating and measuring the impact of energy efficiency policy on energy demand reduction, which is shared internationally and continually improved in the context of a growing interest in energy efficiency policy.

Measuring the multiple benefits

The multiple benefits approach is different from traditional energy efficiency policy, and thus requires different assessment approaches. A key challenge is that many multiple benefits outcomes – whether direct (first-round) or indirect (second-round) – are less tangible and harder to measure or value in a robust and objective manner, and often do not involve "energy" in form or content. As commonly used policy decision-making tools require benefits to be quantified in order to be considered, any un-quantified benefits risk being attributed a value of zero. To date, this has led arguably to the significant undervaluing of the benefits of energy efficiency.

1 See Annex A: Glossary for definitions of these effects.

2 Rebound factors are a form of multiplier calculated on the basis of previously measured rebound effects; they can be used to estimate the rate of rebound that could be expected in a future energy efficiency intervention. See Annex A: Glossary for a further discussion of the pitfalls of factors and multipliers.

© OECD/IEA, 2014.

A range of methods can be used to attribute value to multiple benefit indicators, many of which are already part of standard practice in policy assessment and more generally in the field of economic analysis. These include direct monetisation, indirect monetisation, quantification using proxies, and methods relying on qualitative data. All types of values can be taken into account if the decision-making method used is suitably adapted.

A second challenge is that energy efficiency experts often have no experience in identifying or measuring the types of indicators relevant to multiple benefits. Thus, multiple benefits are sometimes listed but rarely quantified; to date, non-monetary, qualitative approaches to evaluating multiple benefits have been the norm.

This guidebook outlines how available and familiar approaches can be used to measure the multiple benefits of energy efficiency policy and achieve two aims:

- attribute value to all observed impacts, in an effort to overcome the particular challenges of valuing non-market impacts
- estimate and model the wider multiple benefit impacts.

As with measuring energy demand reduction, the impacts selected for analysis will determine which methodology should be used to measure and quantify multiple benefits. Other practical constraints (such as time available, cost restrictions, depth of the data at hand and level of accuracy required) will also influence the choice of method.

Attributing value to all observed impacts

Translating multiple benefit impacts into monetary values is the most effective means of ensuring they are accepted and properly accounted for in policy assessments. When monetisation is not possible, non-monetary quantitative values can be used to factor multiple benefits impacts into the standard methodologies alongside data on kilowatt hours and tonnes of carbon-equivalent saved. Where benefits remain unquantifiable, techniques have been adapted to obtain a clearer qualitative measure of how the impacts of a particular energy efficiency measure are experienced by the beneficiaries.

Most techniques for monetising outcomes fall into place along two spectrums, being either quantitative or qualitative and ranging from subjective to objective. Plotting the techniques – in this case a non-exhaustive list – is useful for seeing which are most suited to the demands of a particular policy assessment (Figure 1).

Quantitative/objective methods

The most robust methods are those which use quantitative values to provide an objective evaluation of the outcome measured, e.g. through **direct linking to a market valuation**. To the greatest extent possible, direct market values specific to each consumer, investor or beneficiary group should be used, rather than applying average prices to all groups. Where a direct equivalent in monetary terms has not yet been determined or is not appropriate, assumptions and corrections can be used to calculate an **indirect market valuation** of outcomes. Examples include: valuing energy reduction directly using prevailing and future consumer prices; valuing labour or physical outputs at current and anticipated market prices; or valuing health benefits at prevailing applicable health services costs.

Case studies from traditional building retrofits demonstrate how to monetise related materials and health benefits. A study carried out in the United States used available data to quantify the value of improving energy efficiency of lighting in a municipal building. In addition to energy cost savings, the study measured savings on ballasts and light fittings based on annualised costs of material replacement (Woodroof et al., 2012). In New Zealand, health improvements driven by residential retrofit measures were valued with reference to

© OECD/IEA, 2014.

the savings arising from reduced days when children stay home from school due to illness. The cost of one sick day was quantified based on the average amount that might be spent on engaging a caregiver to mind the sick child for six hours, applying the minimum wage. This method produced a quantified estimate of the value of each sick day avoided, which was then added to the overall BCA (Preval et al., 2010).

Figure 1 Methods to monetise outcomes

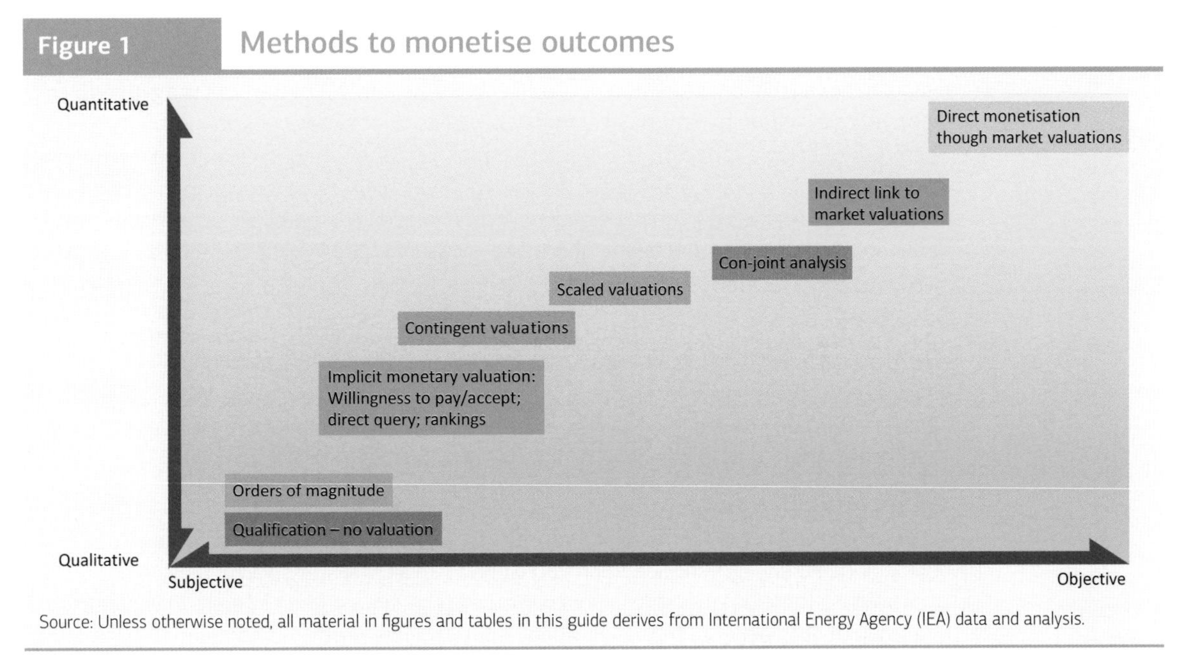

Source: Unless otherwise noted, all material in figures and tables in this guide derives from International Energy Agency (IEA) data and analysis.

Key point *Plotting valuation methods on a two-axis continuum reveals which are more qualitative/subjective and which fit the demands of objective/quantitative results.*

Many multiple benefit indicators reflect social impacts, such as improved comfort or improved brand reputation, and cannot easily be attributed a fixed value in the market. Here different, less objective techniques need to be applied. **Willingness to pay** and **willingness to accept** techniques can, for example, be used to calculate implicit market values with reference to observed consumer choices. Within this method, the **revealed preference** approach can be used to infer an implicit price on a good or service by examining data on consumer behaviour in a real market.

This approach was used in a Massachusetts study of commercial and industrial impacts of energy efficiency measures. Evaluators carried out a large-scale, in-depth interview process, using a series of open-ended questions to identify sources of value for the respondents. They determined values by asking participants to (i) identify any cost and revenue centres impacted, (ii) discern nature of the impact, and (iii) attempt specific valuations of the increase or decrease. Finally, evaluators sought to obtain metrics to measure the magnitude of those changes and to convert time into money (Tetra Tech, Inc. and MPA, 2012).

Quantitative/subjective methods

Subjective reports from recipients of an energy efficiency measure can also be used to monetise values. In what is known as a **stated preference** approach, the quantification derives not from actual market behaviour as above, but from surveying beneficiaries about the value they would place on a certain outcome – essentially, the method characterises a hypothetical choice within a hypothetical market.

© OECD/IEA, 2014.

Conjoint techniques involve asking respondents to rank benefits, then applying econometric techniques to identify the "utility" value of the outcomes, which can then be monetised. For example, another Massachusetts study that sought to measure the "comfort value" customers derived from residential energy efficiency retrofits conducted a survey asking participants to describe impacts and then value them as a fraction of bill savings (or some other observable value), enabling evaluators to derive a contingent valuation based on the energy bill (Tetra Tech, Inc. and MPA, 2011).

Direct query or **choice experiment** methods are even more subjective in that respondents are simply asked to put a value on the benefit. An **orders of magnitude** approach asks participants to make a relative valuation of several possible outcomes, thereby producing a range of values such as minimum or maximum valuation or identifying what cost would support a particular decision or "switching" value. While this method does not elicit concrete values, it has been used to bound early estimates (i.e. establish upper and lower limits) of the magnitude of hard-to-measure multiple benefits, such as improved aesthetics, safety and comfort (Skumatz, 2006).

Efforts to quantify subjective impacts stem from the idea that even rough estimates are more accurate than assuming a value of zero. Simple ranking, for example, can indicate a consumer preference in the absence of any other information. While subjective quantitative methods are open to bias, they provide a concrete measure of value that can be used to inform decision making in diverse contexts.

Qualitative methods

Some multiple benefits do not lend themselves to quantitative assessment. Hard-to-measure benefits might include improved worker morale, greater social cohesion or improved customer relationships. Increasingly, experts show a keen interest in ensuring that the value of such benefits to the individual and to society – which is sometimes significant – is not overlooked. Where quantification is not practical, experts now advocate for the use of qualitative methods.

Methods for measuring these types of impacts generally involve case studies, focus groups, systemic interviews and surveys that ask programme participants to describe the impacts they experienced. The use of **structured surveys**, informed by previously collected evidence or in-depth interviews that ask programme participants to describe in their own words the impacts they experienced, have proven an important tool. Use of structured surveys is standard practise in the mental health profession, where standardised questionnaires allow evaluators to make comparable assessments of mental health status. Applying surveys borrowed from the health profession in the context of energy efficiency policy evaluation can provide a strong basis for action (Liddell, 2013; Gilbertson, 2013). Where quantification is not practical, experts advocate for the triangulation of evidence from mixed methodologies to ensure that all impacts reported by beneficiaries of an energy efficiency intervention are taken into account.

Qualitative research is used regularly to explore occupant reactions to retrofitted insulation and the ways in which they experience and cope with fuel poverty (Critchley et al., 2007; O'Sullivan, Howden-Chapman and Fougere, 2011). Even when evaluations are not able to quantify impacts, case studies can provide powerful evidence and illustrate the meaningful impact of energy efficiency interventions for energy users. The qualitative results about impacts in the lives of fuel-poor households have driven political ambition to address the issue as much as public awareness and acceptance of the need for a policy intervention. Several IEA governments have been motivated by the strength of qualitative evidence

© OECD/IEA, 2014

to target energy efficiency policies to tackle the challenge of fuel poverty, e.g. Ireland's *Affordable Energy Strategy 2011*, the United Kingdom's *Energy Strategy 2013*, and the *Third Energy Package* of the European Union.

Qualitative assessments may also be appropriate or useful when resources (human, financial or time) may simply not be available to carry out the work involved in obtaining reliable, quantified values for multiple benefits of energy efficiency.

Estimating the wider multiple benefits

Valuing the wider, second-round (indirect) impacts of energy efficiency measures presents additional challenges. Yet efforts to make such values available for integration into analytical and decision-making processes are important, in part because the second-round impacts reflect trickle-through effects arising from direct impacts and thus represent additional, cumulative value. Objectively or subjectively attributed values for the full range of impacts can serve as inputs to further analysis of more complex multiple benefit outcomes. The trickle-through effects often flow beyond the stated objectives of an intervention to the wider economy and manifest in various effects such as employment effects, interactions between sectors (which may involve trade-offs), and price effects at sectoral, national and international levels (Box 1).

Modelling or other estimation techniques are often required to fully assess the complex dynamics among direct and indirect impacts of energy efficiency measures. The scope of the assessment (i.e. whether it covers a specific project, an entire sector or the whole economy) and the complexity of the valuation challenge (with associated data requirements being a key factor) usually determines the most appropriate method. These methods are not mutually exclusive; several may be used in conjunction to assess different aspects of a given question. Plotting some of the most commonly used methods along two axes helps visualise how they might be suited to different situations (Figure 3).

At the simplest level, a range of methods support assessment at the project or company level, where complex dynamics are less relevant to the decisions being made. These methods can also be used in situations when time and financial constraints make it impractical to carry out more detailed assessments or when direct empirical data are not available. When the dynamics of indirect impacts are more complex, various modelling techniques can facilitate a more detailed assessment. In general, models can be characterised as either top-down, if they rely on aggregate economic data to assess broad economic effects, or bottom-up, if they use disaggregate engineering or statistical data to look at more specific effects in a more narrow context (e.g. within one sector). Models can also be categorised by whether they focus on price, supply and demand interaction within in a single market or sector or for a single good (partial models), or whether they seek to account for interactions among different sectors/markets and among different goods (whole-economy models) (Ghersi and Hourcade, 2006; Oeko-Institut et al., 2012).

Simple assessment

A first estimate of the impacts of energy efficiency policy can often be carried out using basic calculation methods, such as estimating the direct costs and benefits using simple spreadsheet software. A basic calculation usually begins with a simple equation representing the main investment costs relating to the policy with the expected energy cost reduction for reference and policy scenarios, including the implications for other sectors if possible (Oeko-Institut et al., 2012). The difference between the scenarios should show the estimated impacts of the policy.

© OECD/IEA, 2014.

Box 1 Trickle-through effects in the multiple benefits of energy efficiency

Benefits from energy efficiency improvements can simultaneously manifest in various sectors and across multiple levels; thus, it is useful to consider such impacts separately at each point they arise, as well as the interactions among them. This type of assessment is also extremely important to avoid double-counting impacts when carrying out final benefit-cost evaluations of measures or programmes as a whole. A schematic of the types of effects that could be triggered by energy efficiency retrofitting of a building provide one example of how the direct and indirect impacts might arise (Figure 2).

Figure 2 Simultaneous impacts of a building retrofit programme across sectors and economic levels

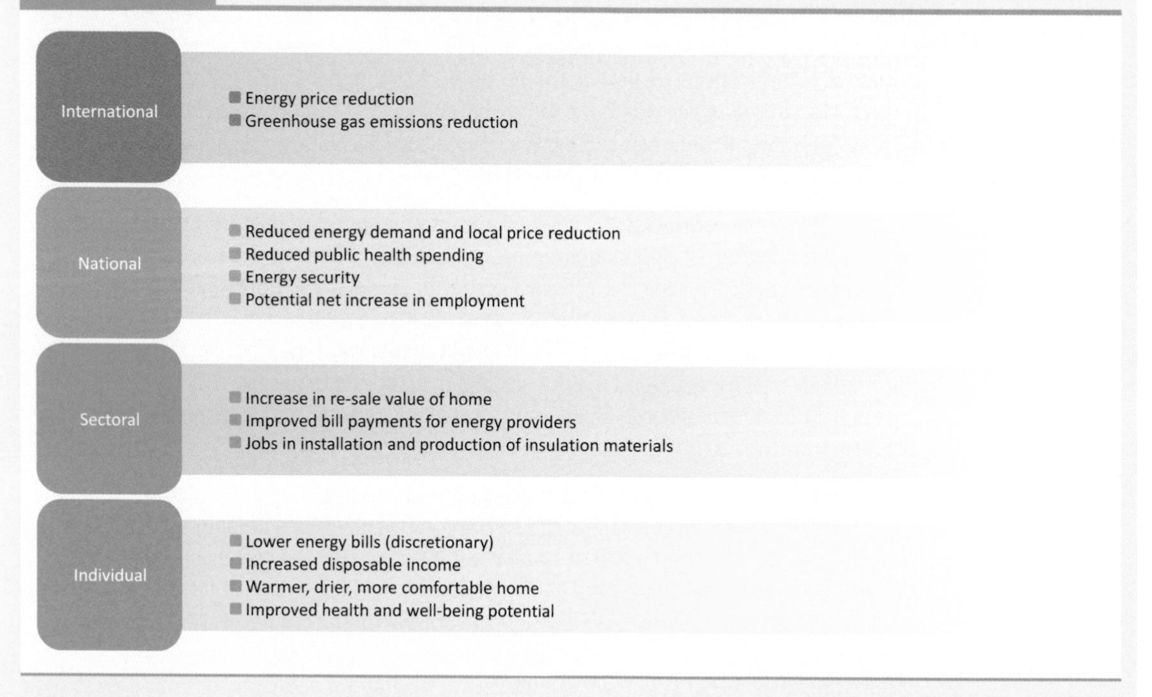

Key point *Benefits can manifest in several areas simultaneously and have trickle-through effects across several areas that must be accounted for in evaluations to avoid double-counting.*

Source: IEA (2012), *Spreading the Net: The Multiple Benefits of Energy Efficiency Improvements*, Insights Paper, OECD/IEA, Paris.

Indirect costs and benefits are difficult to estimate in such a basic assessment, as there are no feedback mechanisms to allow inclusion of effects induced in other parts of the economy or of rebound effects.

- **Activity and fiscal multipliers (or "adders")** are useful tools that can be used in conjunction with measured values to estimate those that have not been measured.[3] This often includes indirect impacts in specific situations or at the economy-wide level. Multipliers are inaccurate and therefore only useful for initial estimates, but can provide indications of areas that might warrant more detailed analysis of policy impacts. Often,

3 See Annex A: Glossary for a more detailed definition.

© OECD/IEA, 2014.

multipliers are based on factors or indices deriving from historical data and may include activity and labour multipliers, fiscal multipliers, estimation of orders of magnitude, or intra-country comparison of existing results.

Figure 3 Methods to assess indirect impacts, accounting for scope and complexity

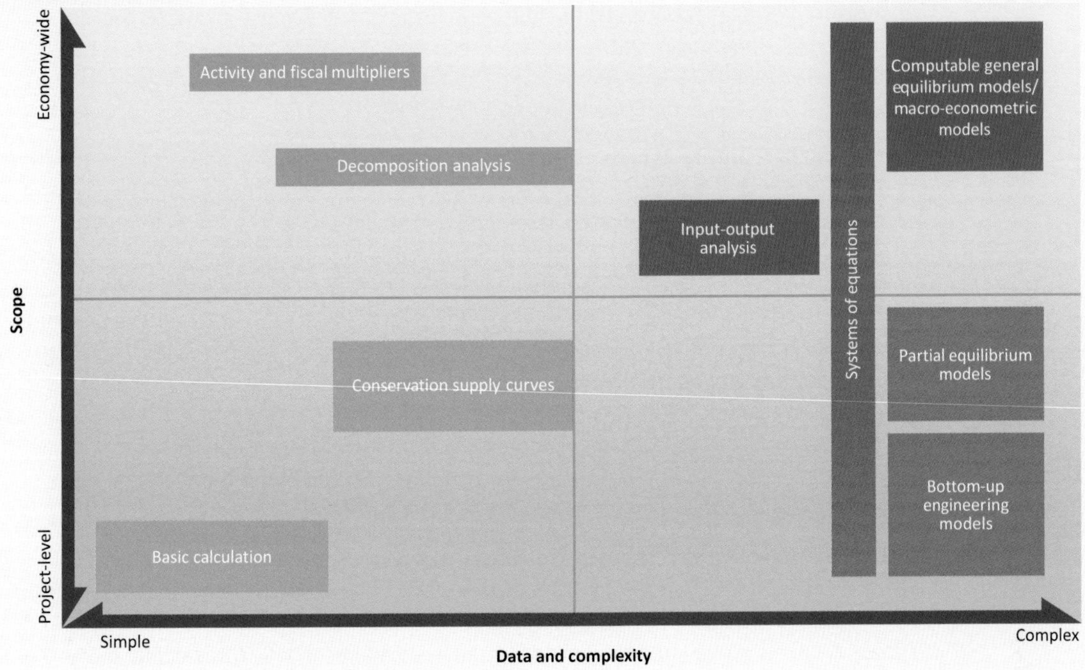

Key point *Modelling methods tend to be best suited to particular applications, but can be used in conjunction to develop a more complete and accurate picture of overall impacts.*

The economic analysis of impacts of an Irish home retrofit grant scheme, for example, applied multipliers in a second-level assessment. Initially, the total annual investment in home retrofits induced by the grant scheme was divided by the average industrial wage in the construction sector to calculate the number of direct jobs supported by the scheme. A multiplier of 1.4 was then applied to reflect the indirect jobs in order to calculate the total number of jobs supported (Scheer and Motherway, 2011).

Governments typically publish fiscal multipliers to reflect the ratio between government spending and gross domestic product (GDP) increase. These figures can be useful in estimating the GDP impacts of a national-level energy efficiency policy. Care is needed in their use, however, as fiscal multipliers differ across and within countries because the structure and behaviour of economies vary (Barrell, Holland and Hurst, 2012).

In the United States, when energy providers report to the regulator on outcomes of their energy efficiency programmes, multiplier factors are commonly used to account for unmeasured multiple benefits accruing to customers. The reported multipliers typically range from 10% to 15% but empirical studies show they should be closer to 45%, or as

© OECD/IEA, 2014.

high as 75%, to reflect the true outcomes of energy efficiency programmes targeting low-income customers[4] (Synapse Energy Economics, Inc., 2014).

- **Decomposition analysis** identifies the socio-economic forces that shape the state of energy-activity intensities in an economy (Ang, 2012). These top-down models usefully discern structural, fuel-mix and exogenous drivers from activity and energy efficiency drivers of energy demand. Decomposition analysis can be particularly useful for policy makers striving to understand the interplay between the environment and socio-economic activities, and to design policies that influence these drivers. An example of this method is the policy analysis framework developed by the European Environment Agency known as the driving forces – pressure – state – impact – response framework (which, in turn, is based on the well-known "PSR" (pressure – state – response) model used by the Organisation for Economic Co-operation and Development (EEA, 2006).

- **Conservation supply curves (CSCs)** were developed to describe and compare different options for energy conservation in a transparent way. They show the quantity and cost of conserved energy, as well as the costs related to specific saving options. In the context of a particular sector or economy, CSCs can thus provide a merit order of most cost-effective options within a common BCA. Together with calculated costs of conserved energy (CCE), bottom-up energy CSCs have been used in early efforts to estimate the multiple benefits impacts for energy efficiency measures in the industrial sector. The net financial savings measured varies greatly from study to study, ranging from 0.03% to 70% of the total savings (Worrell et al., 2003; Hasanbeigi, Menke and Therdyothin, 2010). CSCs can also present the results of bottom-up engineering models by giving an overview of the range of demand-side policy options and their costs compared with the long-run marginal costs of competing supply options.

Complex assessment/economy-wide scope

When an economy-wide assessment is desired but in less detail, top-down models (which rely on aggregate economic data) are often effective. As such models are often used in conventional economic modelling for policy analysis and economic forecasting, many energy efficiency experts are already familiar with them. Their key advantage is that they can represent multiple sectors and model a wide range of interactions. Because they are usually represented as linear, however, they do not provide much detail on the mechanisms that stimulate outcomes. Specific models are appropriate for different assessment aims (Box 3). In general, such models tend to be more onerous in terms of data requirements than basic assessment methods.

- **CGE models** can identify subtle linkages among main sectors of the economy; this top-down assessment allows investigation of the relationships between key sectors, polices, consumers and the government. Many policy analysts use CGE assessments to model macroeconomic impacts because of their capacity to highlight dynamic interactions and effects that can occur over the longer term to affect a variety of macroeconomic indicators. That said, CGE can be limited by its reliance on aggregate assumptions and on the assumption that economic equilibrium is always achieved. New CGE approaches are starting to accommodate imperfect markets. In the north-eastern region of Canada, the *Regional Economic Models, Inc.* (REMI) CGE model has been developed to examine the total net effects (e.g. GDP, employment, energy demand reduction and tax revenue) of investing in all cost-effective efficiency measures for electricity, natural gas and liquid fossil fuels in four provinces. The model uses data on net savings and reduced operating costs

4 These figures were based on publicly filed 2012 Annual Report data for one of the largest energy efficiency programme administrators in Massachusetts.

© OECD/IEA, 2014.

to determine the rate of new investment and economic output that this should generate. Results indicate that for every USD 1 invested in energy efficiency measures, between USD 4 and USD 8[6] of GDP would be generated (ENE, 2012).

Box 2 A pioneering bottom-up model of net benefits

The Net Benefit Model (NBM) is a spreadsheet-based tool developed by the Energy Efficiency and Conservation Authority in New Zealand to assess and prioritise energy efficiency initiatives in homes.

As well as reduced energy costs, the NBM calculates avoided GHG emissions, improved health and direct employment effects. It estimates indoor temperatures before and after retrofit and draws on building energy modelling software to estimate the energy required to achieve those temperatures. This information is then used to model the scale of potential health benefits.

The NBM allows the maximum variety of housing situations to be evaluated. Inputs into the model reflect the specific characteristics of each house that has been retrofitted including: its location, existing levels of insulation, size and age, existing heat source, and occupancy characteristics. The health benefit valuation is from the perspective of national benefit (rather than that of individual consumers), and is based primarily on Ministry of Health data for the costs of various health conditions.

A BCA conducted as part of the evaluation of the *Warm Up New Zealand: Heat Smart* (*WUNZ: HS*) programme conservatively estimated an ongoing annual benefit in healthcare savings of USD 483 (NZD 563) pert household for retrofitted insulation and USD 4 (NZD 4.64) for improved heating systems (Grimes et al., 2011). In a subset of low- to middle-income households, the benefit increases by 45%.[5] The health benefit data are taken from a retrospective cohort study carried out on the first 46 655 houses to receive retrofit treatment (Telfar-Barnard et al., 2011).

To ascribe a value to the health benefits, the study analysed data from three main sources: the Heat Smart programme, the national census, and the New Zealand Ministry of Health on hospitalisations, pharmaceutical dispensations and mortality. Final results of an independently conducted WUNZ: HS evaluation showed benefits in three key areas:

- an estimated USD 1.2 billion to USD 1.5 billion in benefit over 20 years for insulation, 99% of which comprises health benefits with reduced mortality accounting for 74%
- employment benefits estimated at 300 to 800 new jobs created annually
- energy demand reduction of USD 15.6 million (4% discount rate), which is relatively small but statistically significant.

The overall benefit-cost ratio calculated for the programme is over 4:1 (Grimes et al., 2011) (see Box 4.2).

- **Macro-econometric models** provide economy-wide estimates based on historical relationships, which they assume continue over time. They use econometric analysis as described below but cover the whole economy. This reliance on historical data limits the ability of macro-econometric models to reflect structural change (this can be an issue for all models). Macroeconomic impacts were modelled in the European Union, using the macro-econometric model Energy-Environment-Economy Macro-econometric Model for Europe (E3ME), as part of the impact assessment supporting the Energy Efficiency Directive (EC, 2011).
- **Input-output (I-O) analysis** uses statistical data on financial and activity inflows and outflows in various sectors. Data are compiled on the shifts in spending between sectors and then converted into coefficients. This allows researchers to estimate how increased

5 From USD 483 (NZD 563) to USD 702 (NZD 818).
6 These figures were originally calculated in Canadian dollars.

© OECD/IEA, 2014.

spending in one sector affects other sectors. The main drawback is that I-O analysis is rigid and static, with fixed coefficients and structures for all disaggregated production in a given period. Also, the need for quality primary data is high. Nevertheless, I-O is a relatively straightforward way to obtain economy-wide estimates of the indirect and induced effects of spending for energy efficiency programmes. I-O tables can be used alone or incorporated into more complex models. Better understanding of indirect outcomes would improve the representation of sector inflows. I-O analysis is often incorporated into both macro-econometric and CGE models to provide data on the direct, indirect and induced impacts of medium- and large-sized investments in industry sectors over a specific time span.

- **Systems of equations and econometric analysis** can be used to assess both single and multiple sectors, and are particularly valuable if they can draw on a rich set of outcome information. A system of equations involves solving two or more equations simultaneously to describe a set of interactions. Econometric analysis, the economic term used for regression analysis, is a statistical method that uses historical data to estimate behavioural (and unobservable) relationships that cannot be directly measured. It is attractive in that it uses real-world data and is less theoretical than some other approaches. Econometric techniques can be applied to individual or national/regional level data and therefore can be considered as bottom-up or top-down analysis.

Complex assessment/narrow scope

- **Sectoral or partial models** are effective for constructing detailed relationships between inputs and outputs. They can provide accurate results on how policies directly affect key variables at a disaggregated level (e.g. prices, employment and demand at the sectoral and regional level), but are intrinsically limited to a single sector. Sectoral or partial models can be top-down or bottom-up, or a combination of both.

- **Partial equilibrium modelling** is useful for making a very detailed assessment of an individual sector where interventions are unlikely to have much impact on other sectors (or when only considering direct compliance costs to the regulated sector) (Oeko-Institut et al., 2012). This focus on a single sector requires the clearance of the market in that sector, i.e. supply should equal demand, independent of prices and quantities in other sectors. This narrow focus makes partial equilibrium analysis much simpler than general equilibrium modelling (which covers an entire economy). The IEA World Energy Model, which provides estimates for the IEA *World Energy Outlook* each year, uses multiple partial equilibrium models, each run separately to estimate demand levels. Demand projections are then aggregated and fed back to the power sector to determine electricity prices; in turn, prices are fed back to the modules and the models are run iteratively until equilibrium is reached. A similar procedure is used at the global level for coal/oil/gas demand: prices are initially estimated on the supply side and fed back to demand-side models until equilibrium is reached to represent the global energy demand.

- **Bottom-up engineering models** are one type of partial equilibrium model that can integrate the characteristics and technologies of a particular sector. They use technical data (e.g. the energy intensity of old and new technologies, stock data and market data) to calculate expected impacts. In reality, they require significant physical data and costs of technologies. Like other partial models, they usually take demand as an exogenous input and then consider the different ways in which this demand can be met. A key advantage of this method is that it is much better equipped to take into account threshold effects and non-linear relationships (Box 2 above). If, for example, energy efficiency measures caused a significant change to energy prices, a bottom-up model could switch the technology used in the sector once a price threshold had been reached (Oeko-Institut et al., 2012). A bottom-up model can also provide insight into how targets can be reached, and help to identify the least-cost mix of technologies for a given energy or emissions target.

© OECD/IEA, 2014.

Box 3 Choosing the most appropriate assessment method

Considering the range of simple and complex methods available for estimating impacts of energy efficiency policies, choosing the right one for any given task is no small challenge. Several factors should be taken into account, including the time and resources available for the analysis, the quality of data needed and available, and the time frame for the impact estimation.

Working through the following questions and using the decision tree below (Figure 4) can help policy makers decide which method(s) to apply.

- Are indirect impacts from the energy efficiency measures expected, i.e. is more than one economic sector affected? If yes, a general economic model will be required.
- What data can be obtained in a short time frame and what quality is it? General equilibrium, macro-econometric and bottom-up models are data-intensive. If disaggregated data are not available, then I-O or more basic assessment models should be used.
- What resources (staff capacity and expertise, financial resources, time) are available for modelling? CGE and macro-econometric models require significantly more resources than I-O models.
- What timeframe should be modelled? For very long-term forecasts, CGE models tend to be used. Macro-econometric models tend to be better for medium-term rather than long- or short-term effects. Simple assessments should only be used for short-term effects.
- How many different policy scenarios should be compared? More scenarios require longer modelling time and increase the complexity of models; if resources are limited, simpler models (such as BCA) may be best.

Figure 4 Decision tree

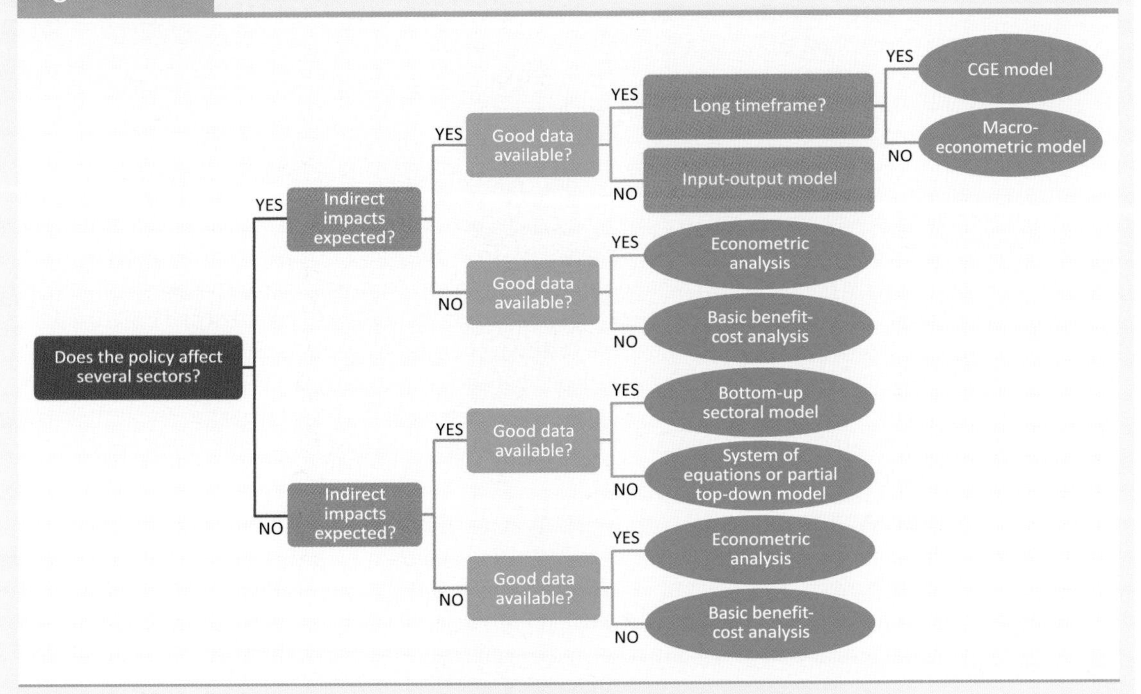

Key point *Several factors should be considered when deciding which assessment method to use.*

© OECD/IEA, 2014.

Bottom-up engineering models have two main drawbacks: they involve the construction of a complex model that can then only be applied to one sector, and a high level of resources is required to support their data needs.[7] They are, however, often useful in combination with other models. A bottom-up engineering model of the buildings sector, for example, can be combined with I-O tables to estimate the effect of energy efficiency measures in the buildings sector on the wider economy (Kronenberg, Kuckshinrichs and Hansen, 2012).

- **Agent models** are a relatively new method; they are bottom-up in design and used to simulate interactions between individual groups (the agents). Agent-based models require significant computer power; as capacity increases in this area, current agent-based models are likely to become more appropriate tools to use for cost assessments.

Using multiple benefits values for policy decision making

Once the impacts in multiple benefits areas have been valued and estimated, they can be integrated into the policy decision-making process, providing policy makers with the means to assess the relative merits of different options and decide which policies to implement, renew or end. Here again, various analytical methods can facilitate the decision-making process, with specific tools being better suited to particular situations (Browne and Ryan, 2011). Because of the cross-disciplinary nature of the outcomes it delivers, the multiple benefits approach requires decision-making tools that can cope with dynamic interactions among sectors and sub-sectors in economic or social systems. The links between energy efficiency measures and their direct and indirect effects can form a complex system, involving interactions among multiple variables – all of which change over time in a pattern of cause and effect (Richardson, 2011). Some traditionally reliable decision-making tools have shown good adaptability for use in assessing multiple benefit impacts, including BCA. Others, particularly when used alone, do not sufficiently address the questions asked in the multiple benefits approach, for reasons explained below. This is true of, for example, cost-effectiveness analysis (CEA). Richer approaches, such as MCDA, are better suited to capturing the potential multiple benefits of a policy in complex systems of causes and effects.

- **CEA** is typically used to assess the capability of different policy options to deliver on very specific stated targets. It is designed to assess the range of costs involved in achieving a single priority benefit and tends to take a narrow view, based only on the cost of the particular programme or policy. This enables policy makers to compare and evaluate the estimated costs of policy options to attain one specific goal; however, it cannot be used to compare the full range of costs and benefits, and thus denies the premise of a multiple benefits approach, which calls for a clear understanding of the full range of outcomes being generated. Recently, some effort has been made to adapt CEA to the multiple benefits policy-making process (Box 4).
- **BCA**[8] is, at present, the most widely used tool for evaluating policy programmes and capital expenditure. It provides a more robust approach for balancing costs against benefits while acknowledging that, while policies should be as cost-effective as possible, the main measure of success should be whether they achieve their objective(s). Ideally, a BCA framework enables the evaluator to account separately for all costs and benefits to both

7 For example, surveys may be needed to obtain the detailed disaggregate information needed to construct the model.
8 Although the commonly used term is "cost-benefit analysis" (CBA), the IEA prefers the term "benefit-cost analysis" (BCA) due to the fact that the ratios produced are expressed as "benefit:cost". The actual approach is the same.

© OECD/IEA, 2014.

government and private sector players. In theory, BCA is thus well-suited to evaluate the multiple benefits of energy efficiency. The drawback is that BCA assumes that all benefits can be monetised, which is not always possible in a multiple benefits context.

Box 4 Five cost-effectiveness tests expanding to include multiple benefits assessment

In the United States, a significant proportion of energy efficiency measures are delivered by energy providers. To support evaluation of these efforts, the Department of Energy and the Environmental Protection Agency proposes five complementary cost-effectiveness tests. Each test represents a different perspective (programme participant, rate-payer, utility, society and total resource cost perspectives), while all specify cost-effectiveness as a primary resource-planning principle (US EPA, 2007).

Standard Practise Guidelines recommend using all five tests together to obtain the most comprehensive information on the impact of a policy; in reality, just one of the tests is generally applied in each assessment. To date, the Total Resource Cost (TRC) Test has been applied most extensively to assess the cost-effectiveness of energy efficiency programmes delivered by energy providers. The TRC is designed to include a range of costs and all benefits incurred by all customers (participating and non-participating) of the utility; however, it provides limited scope to capture harder-to-measure benefits. The SCT has proven more helpful as it covers a broader range of benefits: including benefits for society as a whole, integrating societal discount rates and allowing for consideration of some non-monetised externalities (such as cleaner air and health impacts).

The question of how to address the most pressing challenge facing evaluators today – i.e. integrating multiple benefits into all of these tests – has been left to interpretation, meaning that their use is often inconsistent and results may be inaccurate, fuelling reluctance among regulators to consider these impacts.

Despite these challenges, some US states have forged ahead with incorporating non-energy benefits into cost-effectiveness tests; the evolution underway within the US context is discussed in more detail in Chapter 6.

Most BCA draws on lifecycle cost assessment (LCA) or net present value (NPV) techniques (Box 5) with discount factors to account for the change in real value of investments and outcomes over time. If they can be monetised, the multiple benefits could also be included in techniques such as basic project payback and internal rate of return (IRR) analyses as a way of comparing the benefits and costs. Sometimes, when data are more limited, a partial BCA is carried out and supplemented with qualitative assessment of the other benefits and costs.

To date, most BCA tends to draw on limited assessment of energy demand reduction outputs. Few, if any, of the multiple benefits identified in this publication are currently included in BCA carried out in IEA member countries.

- **Simple payback period** is the length of time required to recover the cost of an investment. It is calculated by comparing the cost of an individual project (e.g. to a company or household) against the cash inflows it generates, and provides an estimate of how long it would take to fully recoup the up-front cost. These inflows tend to be based simply on reduced energy costs resulting from an energy efficiency intervention. Payback period calculations are generally used at the individual project level; they provide a simple metric to assess whether or not to undertake a project or investment – the longer the payback period, the less desirable the investment typically appears. Including other benefits (besides reduced energy costs) in investment calculations should shorten payback periods.

© OECD/IEA, 2014.

Experience with energy efficiency projects in industrial processes shows that a four-year payback can be reduced to two years when broader benefits are integrated into the overall assessment (Lung et al., 2005). All inputs into payback period calculations need to be monetised, which can be a challenge in a multiple benefits context. A weakness of this method is that it ignores the ongoing impacts of interventions, making it necessary to use another method to take into account longer-term costs and benefits.

Box 5 Calculation methods in benefit-cost analysis

Several methods are available to calculate the overall financial value of energy efficiency improvements so that values can be integrated into benefit-cost analysis. Three of the most common are included here.

Lifecycle cost assessment (LCA) calculates the sum of present values of all costs, including investment, capital, installation, energy, operation, maintenance and disposal over the life-time of the project, product, process, programme or measure. It also calculates the present values of any identifiable benefits that arise. LCA is holistic in that it includes the impacts of the energy efficiency intervention on the complete system around a product or process, rather than looking only at specific components. LCA aligns well with the more holistic multiple benefits approach; it could provide a mechanism for integrating multiple benefits impacts into BCA and allow estimation of the full value of a given measure, project or programme over its lifetime.

Net present value (NPV) evaluates the overall current value of a series of cash flows (i.e. related to a project), including all future cash flows. This method requires quantified values of the initial costs, of the costs and benefits for the duration of the calculation period, and some basic economic equations taking into account inflation and depreciation rates over time. If multiple benefits can be translated into cash flows, they can then be integrated into NPV calculations with the likely effect of increasing value. NPV has been used to quantify health and well-being benefits recorded as outcomes of a residential insulation and heating system retrofit programme in New Zealand: the NPV of the avoided cost of hospital admissions was integrated into a BCA framework (Chapman et al., 2009; Preval et al., 2010).

Economic rate of return (ERR)/internal rate of return (IRR) calculations measure the rate of growth a particular project is expected to generate, e.g. for a company or project implementing energy efficiency measures. It is effectively the rate of return to deliver an NPV equal to zero. ERR/IRR supports comparison of the expected value arising from a range of different projects. Again, this is a monetary calculation that requires the multiple benefits outcomes to be monetised.

- **MCDA** supports decision making in the presence of multiple objectives by facilitating comparison of various scenarios within a single framework. It can manage a mixture of quantitative, monetary and qualitative data, as well as varying degrees of certainty. In energy planning, the need to incorporate environmental and social considerations resulted in increased use of multi-criteria approaches. It is commonly used in renewable energy planning to assist with energy resource allocation by assessing various distributed renewable energy options.

  MCDA has proven effective in transportation systems decisions, where it can evaluate alternative strategies, such as eliminating polluting vehicles and choosing between private and public transport in light of a high concern for socio-economic impacts. In Delhi, MCDA

© OECD/IEA, 2014.

was used to examine the impact of including various qualitative criteria in the selection of transportation options (four-stroke two-wheelers, cars and buses operating on compressed natural gas [CNG]). The options were prioritised based on both quantitative criteria (energy demand reduction potential, emission reduction potential, cost of operation) and qualitative criteria (availability of technology, adaptability of the option and barriers to implementation) with both giving different outcomes which could then be considered by the policy maker. Based on quantitative criteria, the CNG cars showed more potential in contributing to an environmentally sustainable transport system in Delhi. By contrast, a combined approach revealed the highest priority for CNG buses followed by four-stroke two-wheelers, CNG cars being the lowest priority (Yedla and Shreshtha, 2003).

This method is well adapted to a multiple benefits approach as it recognises the multi-dimensional nature of sustainability and social issues that should be considered in energy efficiency policy. It provides a transparent presentation of the issues and potential distributional issues trade-offs – a valuable benefit over other approaches, such as BCA. This approach can include elements of subjectivity, especially in the weighting stage at which analysts need to assign relative importance to the criteria. Recognising that the mix of different types of data makes it difficult to show clearly whether benefits outweigh costs, MCDA is designed to facilitate compromise and collective decisions. The European Commission recommends MCDA as a highly flexible approach that could be applied to assess multiple benefits (EC, 2009).

Some innovative approaches that respond to these complexities are beginning to emerge; more can be expected as the multiple benefits approach gains traction (Box 6).

Box 6 A promising option: System dynamics modelling

Current policy approaches tend to focus on one objective (e.g. energy demand reduction) and then attempt to minimise "unintended consequences". System dynamics modelling (SDM) facilitates a more complete understanding of complex causal feedback loops driving the effects of proposed policies over time. It enables a dynamic simulation of policy options and their consequences, identification of realistic targets and trade-offs, and a reflective assessment of the comparative magnitude of initiatives to optimise shared outcomes. When used in a participatory manner, SDM facilitates collaborative learning about the policy system and can build consensus about the most effective policy levers for optimising shared objectives and minimise negative ones (van den Belt, 2004; Beall and Ford, 2010).

This method has been used to improve decision making across various disciplines. Experts at University College London (UCL) are currently applying SDM to develop a shared "causal theory" around several variables, including housing, energy and well-being around seven themes: community connection and the physical quality of neighbourhoods; energy efficiency and climate change; fuel poverty and indoor temperature; household crowding; housing affordability; land ownership, value and development patterns; and air quality and ventilation. These are all deeply interconnected (Figure 5).

This integrated decision-making approach can be used to inform and improve BCA, and can also be used in conjunction with MCDA as part of an iterative policy process.

© OECD/IEA, 2014.

Figure 5 Schematic of interactions in a system comprising housing, energy and well-being

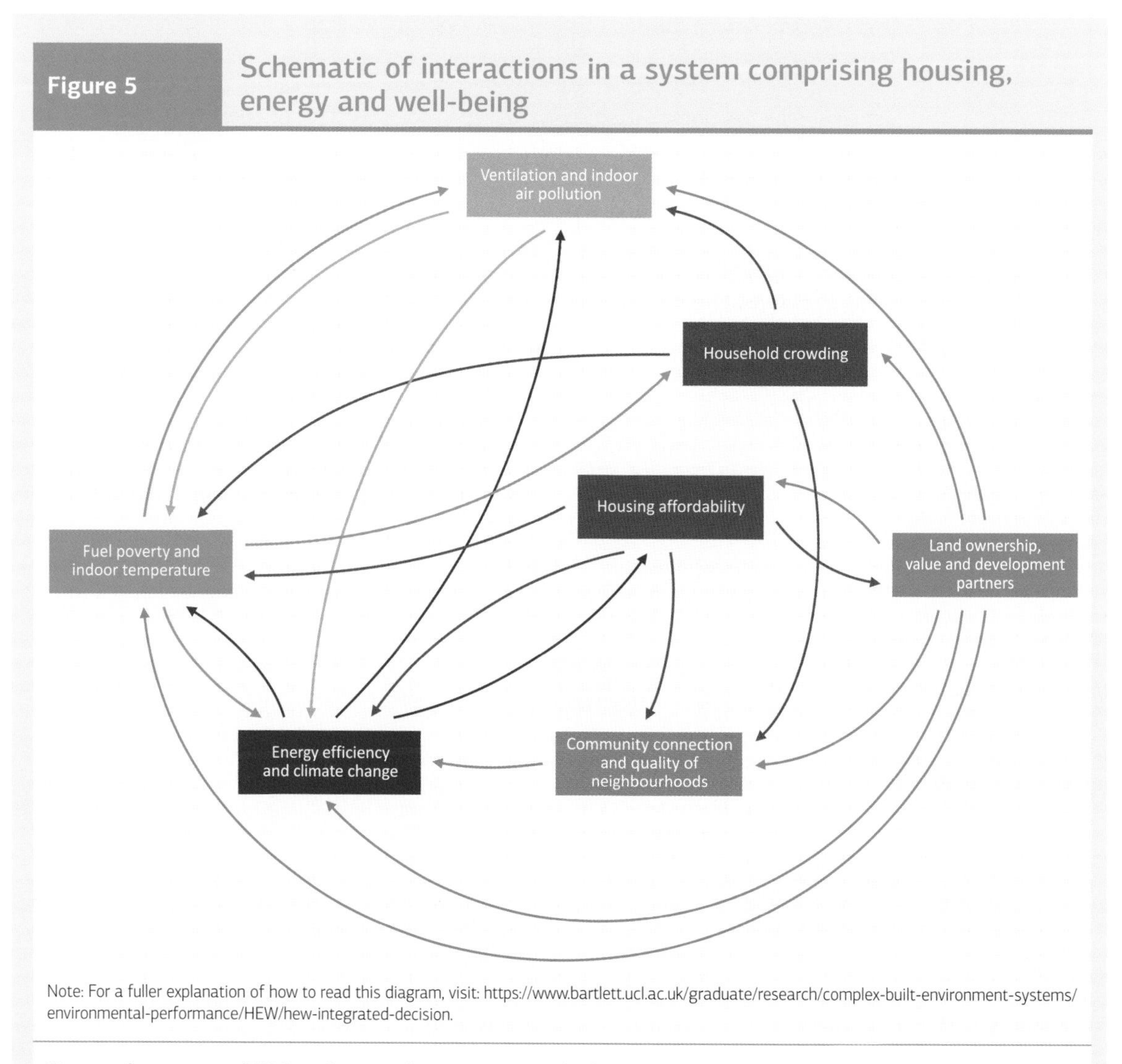

Note: For a fuller explanation of how to read this diagram, visit: https://www.bartlett.ucl.ac.uk/graduate/research/complex-built-environment-systems/environmental-performance/HEW/hew-integrated-decision.

Key point *SDM explores policy options and identifies those which optimise shared goals.*

Source: Macmillan, A., M. Davies and Y. Bobrova (2014), *Integrated Decision-making about Housing, Energy and Wellbeing (HEW), Report on the Mapping Work for Stakeholders*, UCL Complex Built Environment Systems, London.

Integrating multiple benefits into the policy process

A multiple benefits approach requires rigorous effort in the areas of gathering data about impacts, measuring the benefits and taking a holistic approach to policy decision making – all of which have implications for all phases of the policy process. The following section highlights how a multiple benefits approach can be integrated into an IEA Policy Pathway,[9] outlining which additional actions might be taken at each of the four stages: plan, implement, monitor and evaluate (Table 1) to ensure the full range of impacts are taken into account.

9 www.iea.org/publications/policypathwaysseries/.

© OECD/IEA, 2014.

Table 1 Integrating the multiple benefits approach in an IEA Policy Pathway

Policy pathway	Essential steps	Multiple benefits approach
PLAN: The benefits targeted will have direct implications for policy design and implementation planning. Programmes can be designed to maximise the prioritised benefits while minimising both costs and negative impacts.		
Identify the problem requiring intervention	Set prioritised objectives. Identify relevant policy options/responses. Estimate the economic costs and benefits of each option. Identify potential barriers.	Consider which benefits to assess. Consider data availability for each. Choose estimation method best suited to the nature of the selected benefits. Establish baselines for all benefit indicators. Estimate positive/negative impacts on indicators. Integrate values and consideration of policy interactions into chosen assessment framework.
Engage stakeholders early	Identify key stakeholders. Consult stakeholders. Explore co-operation opportunities.	Involve multidisciplinary stakeholders with expertise and interest in relevant benefit areas.
Establish policy framework and action plan	Develop action plan. Prepare contingency plans. Determine how to measure progress.	Determine best-fit methodology for measuring progress in achieving multiple benefits.
Secure resources	Investigate options for resourcing project (including technical, institutional and human elements).	Market the project in terms of multiple benefits. Approach range of funders interested in different benefits for different reasons.
IMPLEMENT: A multiple benefits approach typically aims to take into account a broader range of issues during implementation. It is important to ensure that implementers are properly trained in multiple benefits.		
Engage actors and begin implementation	Call for tenders. Confirm roles, responsibilities, outputs. Launch policy.	Engage stakeholders who can contribute multiple benefits implementation expertise (e.g. healthcare professionals, community groups, industrial engineers, process managers, economists, power producers).
Raise awareness	Communicate targets and goals to implementers and target group. Develop training materials and provide training to implementers.	Recruit implementers with experience in targeted outcomes. Inform implementers and target group about the multiple benefits during implementation.
Manage implementation process	Verify progress, ensure compliance and enforce deliverables. Build capacity and project support.	Include training on multiple benefits.
MONITOR: A multiple benefits approach often introduces new indicators in the list of criteria to be monitored, which is dictated by the outcomes targeted in the policy planning stages. A broad range of relevant experts beyond the energy efficiency discipline should be engaged and innovative data sources utilised.		
Match data collection and analysis to priorities	Set clear data goals and define assessment methodology. Decide which data to collect and from whom. Determine how to analyse results. Publish monitoring data.	Select, monitor and publish data related to the multiple benefits. Engage evaluators experienced in the relevant outcome areas. Give additional attention to qualitative data.

© OECD/IEA, 2014.

Policy pathway	Essential steps	Multiple benefits approach
EVALUATE: Retrospective analysis at the conclusion or revision of a policy intervention is of utmost importance to a multiple benefits approach. Robust evidence is needed to better inform future energy policy decisions.		
Evaluate effects of policy	Analyse data. Assess policy results. Decide who has access to data.	Engage experts from relevant disciplines in the assessment of data. Consider use of innovative data sources.
Communicate results	Establish communication channels. Involve a variety of stakeholders in the communication of results.	Consider whether outcomes are positive or negative from multiple points of view. Present results according to different policy objectives, noting trade-offs.
Report lessons learned	Adapt policies with regards to results. Plan next steps and future actions.	Ensure results are disseminated to all ministries with a potential stake in the outcomes. Facilitate prioritisation of benefits for the future.

Key point *A multiple benefits approach has implications for all phases of the policy process.*

Conclusions

The novelty of multiple benefits presents some methodological challenges. This guide demonstrates how currently available tools can be used to overcome those challenges and put the multiple benefits approach into practice now. In some cases, this means that qualitative and simple quantitative approaches may be most appropriate, and the capacity of these methods to support policy decision making should not be underestimated. More holistic decision-making methods, such as MCDA, enable the range of outcomes of an energy efficiency policy to be considered together and the various trade-offs to be carefully managed.

Policy assessment guidelines are, by nature, intended to be adapted to an evolving policy context and, indeed, evaluations carried out in several IEA member countries to date have been successful in adapting more traditional methods to generate meaningful results about the impact of multiple benefits on different sectors of the economy and society. Through increased usage, existing policy assessment tools are expected to develop and new ones will emerge to better serve the growing interest of policy makers in the multiple benefits that energy efficiency delivers. That development will enable multiple benefits to be more robustly integrated throughout the energy efficiency policy process.

Bibliography

Ang, B.W. (2012), *A Simple Guide to LMDI Analysis*, Department of Industrial and Systems Engineering National University of Singapore, Singapore, www.ise.nus.edu.sg/staff/angbw/pdf/A_Simple_Guide_to_LMDI.pdf (accessed 28 June 2014).

Barrell, R., D. Holland and I. Hurst (2012), "Fiscal Consolidation: Part 2. Fiscal Multipliers and Fiscal Consolidations", *OECD (Organisation for Economic Co-operation and Development) Economics Department Working Papers*, No. 933, OECD Publishing, Paris, http://dx.doi.org/10.1787/5k9fdf6bs78r-en (accessed 28 June 2014).

Beall, A. and A. Ford (2010), "Reports from the field: Assessing the art and science of participatory environmental modeling", *International Journal of Information Systems and Social Change*, Vol. 1, No. 2, IGI Publishing Hershey, Hershey, pp. 72-89, www.igi-global.com/article/reports-field-assessing-art-science/42116 (accessed 28 June 2014).

© OECD/IEA, 2014.

Browne, D. and L. Ryan (2011), "Comparative analysis of evaluation techniques for transport policies", *Environmental Impact Assessment Review*, Vol. 31, No. 3, Elsevier Inc., Amsterdam, pp. 226-333, www.sciencedirect.com/science/article/pii/S0195925510001460 (accessed 8 July 2014).

Chapman, R. et al. (2009), "Retrofitting housing with insulation: A cost benefit analysis of a randomised community trial", *Journal of Epidemiology and Community Health*, Vol. 63, No. 4, BMJ Publishing Group Ltd., London, pp. 271-277, www.ncbi.nlm.nih.gov/pubmed/19299400 (accessed 25 June 2014).

Critchley R. et al. (2007), "Living in cold homes after heating improvements: Evidence from Warm-Front, England's Home Energy Efficiency Scheme", *Applied Energy*, Vol. 84, No. 2, Elsevier Ltd., Amsterdam, pp. 147–158, www.sciencedirect.com/science/article/pii/S0306261906000791 (accessed 25 June 2014).

CPUC (California Public Utilities Commission) (2001), *California Standard Practice Manual Economic Analysis of Demand-side Programs and Projects*, Governor's Office of Planning and Research, State of California, Sacramento, http://cleanefficientenergy.org/sites/default/files/07-J_CPUC_STANDARD_PRACTICE_MANUAL.pdf (accessed 28 June 2014).

EC (European Commission) (2011), Commission Staff Working Paper Impact Assessment, Accompanying the document Directive of the European Parliament and of the Council on Energy Efficiency and amending and subsequently repealing Directives 2004/8/EC and 2006/32/EC, SEC(2011) 779 final, Brussels, 22.6.2011, EC, Brussels, http://ec.europa.eu/energy/efficiency/eed/doc/2011_directive/sec_2011_0779_impact_assessment.pdf (accessed 22 June 2014).

EC (2009), *Impact Assessment Guidelines*, Sec(2009) 92, EC, Brussels, http://ec.europa.eu/smart-regulation/impact/commission_guidelines/docs/iag_2009_en.pdf (accessed 22 June 2014).

EC DG ESAI (Directorate-General for Employment, Social Affairs and Inclusion) (2009), *Guidance for Assessing Social Impacts within the Commission Impact Assessment System*, Ares(2009)326974 – 17/11/2009, EC, Brussels, http://ec.europa.eu/smart-regulation/impact/key_docs/docs/guidance_for_assessing_social_impacts.pdf (accessed 22 June 2014).

EEA (European Environment Agency) (2006), *The DPSIR Framework used by the EEA*, EEA, Copenhagen, http://ia2dec.ew.eea.europa.eu/knowledge_base/Frameworks/doc101182 (accessed 28 June 2014).

ENE (Environment Northeast) (2012), Energy Efficiency: Engine of Economic Growth in Eastern Canada: *A Macroeconomic Modeling & Tax Revenue Impact Assessment*, ENEOttawa, www.env-ne.org/public/resources/ENE_EnergyEfficiencyEngineofEconomicGrowth_EasternCanada_EN_2012_0611_FINAL.pdf (accessed 23 June 2014).

Grimes A. et al. (2011), *Cost Benefit Analysis of the Warm Up New Zealand: Heat Smart Program*, report for the Ministry of Economic Development, Motu Economic and Public Policy Research, Wellington, www.healthyhousing.org.nz/wp-content/uploads/2012/05/NZIF_CBA_report-Final-Revised-0612.pdf (revised June 2012, accessed 22 June 2014).

Hasanbeigi, A., C. Menke and A. Therdyothin (2010), "The use of conservation supply curves in energy policy and economic analysis: The case study of Thai cement industry", *Energy Policy*, Vol. 38, No. 1, Elsevier Ltd., Amsterdam, pp. 392-405, www.sciencedirect.com/science/article/pii/S0301421509007137 (accessed 28 June 2014).

Ghersi, F. and J. C. Hourcade (2006), "Macroeconomic consistency issues in E3 modeling: The continued fable of the elephant and the rabbit", *The Energy Journal*, in special issue "Hybrid modeling of energy environment policies: reconciling bottom-up and top-down", International Association of Energy Economics (IAEE), Cleveland, pp. 39-62, www.centre-cired.fr/IMG/pdf/Epreuve_EJ_17oct061.pdf (accessed 23 June 2014).

Gilbertson, J. (2013), "Psychosocial routes from housing investment to health: Evidence from England's home energy efficiency scheme", presentation at the IEA-EEA Roundtable on the Health & Well-being Impacts of Energy Efficiency Improvements, Copenhagen, 18-19 April 2013, www.iea.org/workshop/roundtableonthehealthwell-beingimpactsofenergyefficiencyimprovements.html (accessed 22 June 2014).

IEA (International Energy Agency) (2013a), *World Energy Outlook 2013*, OECD/IEA, Paris, www.iea.org/W/bookshop/add.aspx?id=455 (accessed 28 June 2014).

IEA (2013b), *World Energy Model Documentation*, 2013 Version, OECD/IEA, Paris, www.worldenergyoutlook.org/media/weowebsite/2013/WEM_Documentation_WEO2013.pdf (accessed 8 July 2014).

© OECD/IEA, 2014.

IEA (2012), *Spreading the Net: The Multiple Benefits of Energy Efficiency Improvements*, Insights Paper, OECD/IEA, Paris, www.iea.org/publications/insights/insightpublications/name,26319,en.html (accessed 18 June 2013).

IEA (2010), *Monitoring, Verification and Enforcement: Improving Compliance within Equipment Energy-efficiency Programmes*, Policy Pathways, OECD/IEA, Paris, www.iea.org/publications/freepublications/publication/monitoring-1.pdf (accessed 28 June 2014).

Kronenberg, T., W. Kuckshinrichs and P. Hansen (2012), *Macroeconomic effects of the German Government's Building Rehabilitation Program*, MPRA (Munich Personal RePEc Archive) Paper 38815, University Library of Munich, Germany, http://mpra.ub.uni-muenchen.de/38815/1/MPRA_paper_38815.pdf (accessed 23 June 2014).

Kushler, M., S. Nowak and P. Witte (2012), *A National Survey of State Policies and Practices for the Evaluation of Ratepayer-funded Energy Efficiency Programs*, research report U122, ACEEE (American Council for an Energy-Efficient Economy), Washington DC, www.aceee.org/research-report/u122 (accessed 3 July 2014).

Liddell, C. (2013), "Tackling fuel poverty: Mental health impacts and why these exist", presentation at the IEA-EEA Roundtable on the Health & Well-being Impacts of Energy Efficiency Improvements, Copenhagen, 18-19 April 2013, www.iea.org/workshop/roundtableonthehealthwell-beingimpactsofenergyefficiencyimprovements.html (accessed 22 June 2014).

Lung, R.B. et al. (2005), "Ancillary savings and production benefits in the evaluation of industrial energy efficiency measures", *Proceedings of the 2005 ACEEE Summer Study on Energy Efficiency in Industry*, Vol. 6, West Point, 19-22 July 2014, ACEEE (American Council for an Energy-Efficient Economy), Washington DC, pp. 6-103-6-114, www.aceee.org/files/proceedings/2005/data/index.htm (accessed 22 June 2014).

Macmillan, A., M. Davies and Y. Bobrova (2014), *Integrated Decision-making about Housing, Energy and Wellbeing (HEW)*, report on the mapping work for stakeholders, UCL (University College London) Complex Built Environment Systems, London, www.bartlett.ucl.ac.uk/graduate/research/complex-built-environment-systems/environmental-performance/HEW/hew-integrated-decision/project-outputs (accessed 8 July 2014).

NZ Treasury (New Zealand Government Treasury) (2010), *Cross Agency Initiatives Process Guide for Public Service Agencies*, New Zealand Government, Wellington, www.treasury.govt.nz/publications/guidance/planning/caip, (accessed 16 March 2014).

Oeko-Institut et al. (2012), *Ex-post quantification of the effects and costs of policies and measures*, final report, CLIMA.A.3/SER/2010/0005, Oeko-Institut e. V., Berlin, http://ec.europa.eu/clima/policies/g-gas/monitoring/docs/report_expost_quantification_en.pdf (accessed 14 April 2014)

O'Sullivan, K., P. Howden-Chapman and G. Fougere (2011), "Making the connection: the relationship between fuel poverty, electricity disconnection and prepayment metering", *Energy Policy*, Vol. 39, No. 2, Elsevier Ltd., Amsterdam, pp. 733-741, www.sciencedirect.com/science/article/pii/S0301421510007974 (accessed 28 June 2014).

Pohekar, S.D. and M. Ramachandran (2004), "Application of multi-criteria decision making to sustainable energy planning – A review", *Renewable and Sustainable Energy Reviews*, Vol. 8, No. 4, Elsevier Science Ltd., Amsterdam, pp. 365-381, http://www.sciencedirect.com/science/article/pii/S1364032104000073 (accessed 28 June 2014).

Preval, N. et al. (2010), "Evaluating energy, health and carbon co-benefits from improved domestic space heating: A randomised community trial", *Energy Policy*, Vol. 38, No. 8, Elsevier Ltd., Amsterdam, pp. 3965-3972, www.sciencedirect.com/science/article/pii/S0301421510001837 (accessed 28 June 2014).

Richardson, G. P. (2011), "Reflections on the foundations of system dynamics." *System Dynamics Review*, Vol. 27 No. 3, John Wiley & Sons Ltd., West Sussex, pp. 219-243.

Scheer, J. and B. Motherway (2011), *Economic Analysis of Residential and Small-business Energy Efficiency Improvements*, Sustainable Energy Authority of Ireland, Dublin.

Skumatz, L. (2006), *Measuring "Hard to Measure" Non-energy Benefits (NEBs) from Energy Programs: Methods and Results*, American Evaluation Association, Portland.

Synapse Energy Economics (2014), Unpublished material provided to the IEA, 3 March 2014.

© OECD/IEA, 2014

Telfar-Barnard, L., et al. (2011), *The Impact of Retrofitted Insulation and New Heaters on Health Services Utilisation and Costs, Pharmaceutical Costs and Mortality: Evaluation of Warm Up New Zealand: Heat Smart*, report to the Ministry of Economic Development, MED, Wellington, www.motu.org.nz/ (accessed 21 June 2014).

Tetra Tech, Inc. and Massachusetts Program Administrators (MPA) (2012) *Final Report – Commercial and Industrial Non-energy Impacts Study*, 20 June 2012, Tetra Tech Inc., Madison, http://www.rieermc.ri.gov/documents/evaluationstudies/2012/KEMA_2012_MA_CI_NEI_REPORT.pdf (accessed 17 May 2014).Tetra Tech, Inc. and MPA (Massachusetts Program Administrators) (2011), *Massachusetts Special and Cross-sector Studies Area, Residential and Low-income Non-energy Impacts (NEI) Evaluation*, final report prepared for MPA, Tetra Tech, Inc., Madison, http://www.rieermc.ri.gov/documents/evaluationstudies/2011/Tetra_Tech_and_NMR_2011_MA_Res_and_LI_NEI_Evaluation(76).pdf (accessed 28 June 2014).

Turner, K. (2013), "'Rebound' Effects from Increased Energy efficiency: A Time to Pause and Reflect", *The Energy Journal*, Vol. 34, No. 4, International Association for Energy Economics (IAEE), Cleveland, www.iaee.org/en/publications/ejarticle.aspx?id=2523 (accessed 18 June 2014).

UK DECC (United Kingdom Department of Energy & Climate Change) (2013), *Annual Energy Statement 2013*, presented to Parliament by the Secretary of State for Energy and Climate Change by Command of Her Majesty, United Kingdom government, London, www.gov.uk/government/uploads/system/uploads/attachment_data/file/254250/FINAL_PDF_of_AES_2013_-_accessible_version.pdf (accessed 28 June 2014).

UK HM Treasury (United Kingdom, Her Majesty's Treasury) (2003), *The Green Book: Appraisal and Evaluation in Central Government*, United Kingdom government, London, https://www.gov.uk/government/uploads/system/uploads/attachment_data/file/220541/green_book_complete.pdf (accessed 28 June 2014).

US EPA (Unites States Environmental Protection Agency) (2008), *National Action Plan for Energy Efficiency: Understanding Cost-effectiveness of Energy Efficiency Programs: Best Practices, Technical Methods, and Emerging Issues for Policy-makers*, Energy and Environmental Economics, Inc. and Regulatory Assistance Project, US EPA, Washington DC, www.epa.gov/eeactionplan (accessed 22 June 2014).

US EPA (2007), *National Action Plan for Energy Efficiency: Model Energy Efficiency Program Impact Evaluation Guide*, prepared by Schiller, S.R., Schiller Consulting, Inc., US EPA, Washington DC, www.epa.gov/eeactionplan (accessed 22 June 2014).

van den Belt, M. (2004), *Mediated Modelling. A System Dynamics Approach to Environmental Consensus Building*, Island Press, Washington DC, http://islandpress.org/ip/books/book/islandpress/M/bo3558672.html (accessed 28 June 2014).

Woodroof, E.A. et al. (2012), "Energy conservation also yields: capital, operations, recognition and environmental benefits", *Energy Engineering*, Vol. 109, No. 5, Taylor & Francis, pp. 7-26, www.tandfonline.com/doi/abs/10.1080/01998595.2012.10531820?tab=permissions#tabModule (accessed 28 June 2014).

Worrell, E. et al. (2003), "Productivity benefits of industrial energy efficiency measures", *Energy*, Vol. 28, No. 11, Elsevier Ltd., Amsterdam, pp. 1081-1098, www.sciencedirect.com/science/article/pii/S0360544203000914 (accessed 23 June 2014).

Yedla, S. and R.M. Shreshtha (2003), "Multicriteria approach for selection of alternative options for environmentally sustainable transport system in Delhi", *Transportation Research Part A*, Vol. 37, No. 8 , Elsevier Ltd., Amsterdam, pp. 717-729, www.sciencedirect.com/science/article/pii/S0965856403000272 (accessed 28 June 2014).

© OECD/IEA, 2014.

Annexes

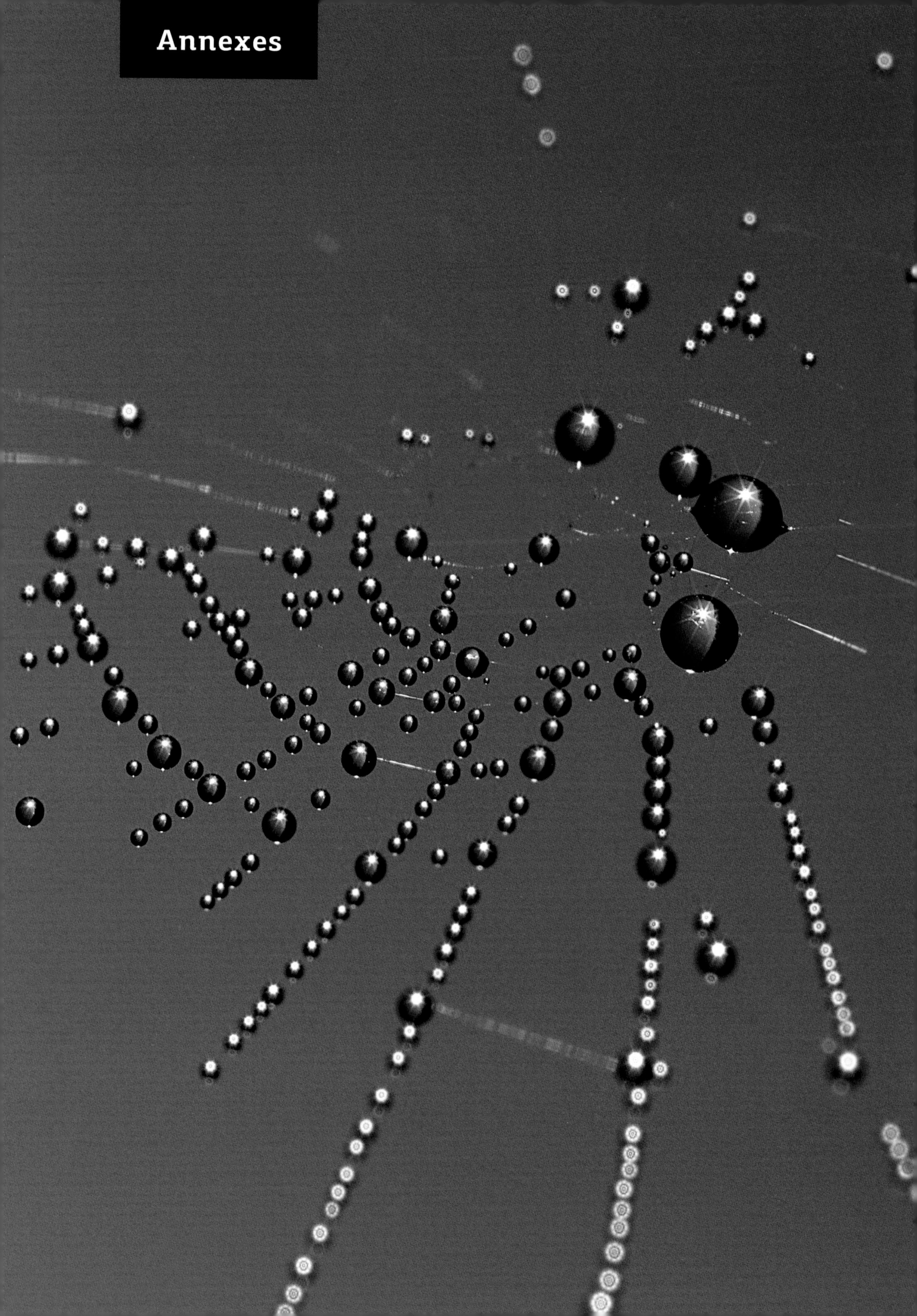

Glossary

This glossary clarifies how key terms have been used in the investigation, by the International Energy Agency, of a multiple benefits approach, and other terms used in this publication.

Adders and multipliers are factors that can be used in conjunction with measured values to estimate impacts that have not been measured, based on existing evidence of that impact. **Adders**, for example, are allowed by United States energy utility regulators to express the additional value that consumers are expected to derive from investments in energy efficiency. **Labour multipliers** indicate the expected increase in upstream and/or downstream labour in a sector or economy resulting from an increase in labour activity at a particular point in the system. **Fiscal multipliers** represent factors that governments may publish as official statistics and can be applied to changes in government spending to estimate a change in gross domestic product and/or tax revenues.

Ancillary benefits/co-benefits have been traditionally used to describe the impacts of energy efficiency beyond reductions in energy demand – i.e. the benefits that occur in addition to a single prioritised policy goal. While these terms have been used interchangeably with multiple benefits in other literature, this publication opts to use multiple benefits in order to avoid a pre-emptive prioritisation of various benefits; different benefits will be of interest to different stakeholders.

Benefit-cost ratio is the ratio of monetised outcome benefits to project investment costs (it can be understood interchangeably with cost-benefit ratio, used in some publications). This calculation is the result of a benefit-cost analysis, a commonly used method for assessing whether a policy delivers good value or return on investment for its actual cost.

Crowding is the degree to which a policy or intervention displaces other inputs or investments in a market (crowding out), or draws in additional inputs or investments (crowding in). Crowding includes the distortion and/or displacement of expected investment flows within and across markets.

Default values are used as estimates in energy efficiency policy assessment in place of measured values; they are frequently based on averages from a large number of measured values. Default values are mainly used when data are scarce for a particular situation or when the technical capacity to carry out a detailed calculation is lacking.

Deferral (network deferral) is the amount of actual or anticipated demand or transmission capacity growth that is delayed or deferred by downstream improvements to system operation and load reduction.

Direct or "first-round" impacts are those impacts or effects arising directly from the energy efficiency measure, such as reduced energy demand or increased investment in energy efficiency-related goods and services.

Discount factor is the ratio applied to current values in order to derive a value for future annual revenues and costs; it reflects factors such as perceived future risk and the premium that is placed on immediate revenues and deferred costs.

© OECD/IEA, 2014.

Effect describes an additional factor (or factors) that can influence how benefits and impacts manifest.

Energy efficiency improvement is an improvement in the ratio of energy consumed to the output produced or service performed. This improvement results in the delivery of more services for the same energy inputs or the same level of services from less energy input.

Energy efficiency measure is any action or activity undertaken with the aim of improving the ratio of energy consumed to the output produced or service performed.

Energy management is the activity within the operation of industrial, commercial and public sector facilities of monitoring, auditing, managing and implementing improvements to systems that demand and consume energy.

Energy provider is an entity engaged in selling or delivering energy to customers; it can include utilities, electricity or gas retailers and distribution businesses, or providers of batch delivered oil, liquefied petroleum gas or solid fuels.

Fuel poverty refers to a situation in which a household technically has access to energy but cannot afford adequate energy services to meet their basic needs (see Box 4.1).

Impact is any kind of result from an action or measure. In this publication, impact is used to describe any result, positive or negative, arising from an energy efficiency measure. In this context, the impact could be reduced energy consumption, for example, or increased economic activity (which may drive up energy consumption overall).

Indicator is an observable or measurable result that shows evidence of whether an impact has occurred and the nature of that impact. It provides a metric by which one can quantify and define the scale of a resulting change.

Indirect or "second-round" impacts refers to impacts or effects from the direct impacts such as increased disposable income resulting from reduced energy consumption and costs. The indirect impact might manifest as increases in spending, employment or gross domestic product.

Induced impacts refer to impacts that arise further down the causal chain, as a result of indirect impacts (see definition above); examples might include additional spending by the people employed as a result of direct or indirect benefits.

Line losses refer to the amount of electricity lost as electricity passes through transmission lines to customers, as a result of the electrical resistance inherent in any transmission and distribution system.

Long-run marginal cost is the cost of providing an incremental additional unit of capacity in the electrical system over the long run, typically expressed in USD per megawatt hour. This includes amortising capital assets and the operation and maintenance costs over the long run. As long-run marginal cost defines the cost of providing the next (marginal) unit of demand, it is the key component of wholesale electricity prices.

Monetisation is the attribution of financial value to phenomena, usually by relating a change in status of a good or service to the relevant market value of the good or service.

Multiplier effect is a further extension of an induced impact, referring to ripple effects arising across the wider economy from the original energy efficiency policy. For example, a multiplier effect would be that stores, restaurants or other service providers benefit from the spending of people who are newly employed (directly or indirectly) because of an energy efficiency policy and have greater capacity to spend or invest their earnings.

© OECD/IEA, 2014.

Multipliers: see definition of adders above.

Net benefit is the measure of the value of an outcome after the cost of delivering the outcome has been accounted for and deducted.

Non-energy benefits (NEBs) is a term that has been commonly used (particularly in the United States) to describe the benefits of energy efficiency beyond traditional energy savings. This concept is distinct from "multiple benefits", which refers to *all* benefits arising from energy efficiency, including traditional energy-related benefits and benefits in other areas.

Non-market impacts describes benefits arising from energy efficiency measures that cannot be directly linked to market values.

Peaker (equivalent peaker) refers to an energy generation installation (or plant) used only to meet peak demand. An "equivalent peaker" describes a demand-side initiative that can be triggered to offset a demand peak; for example, a supply contract with businesses to shut off shed-able load during demand peaks.

Public procurement is the process by which governments (at national, regional or municipal levels) purchase products and service used in the provision of public services.

Rate-payer subsidies are financial contributions charged to individual energy consumers on their energy bills, imposed by energy system regulators to subsidise the costs of building new power plants or other improvements to the energy system that are expected to benefit all energy users.

Rebound effect see definition set out in Chapter 1.4 of this publication.

Revenue-neutral reflects the provision of products and services by the public sector in such a way that costs are offset with revenues or reduced costs, so that there is no net charge against public budgets.

Spill-over effects are externalities of economic activity or processes that affect individuals or other entities not directly involved. For example, in the case of energy efficiency, residents living near a polluting factory that reduces energy consumption – and thus emissions – would enjoy the spill-over effect of cleaner air.

Weatherisation refers to a package of energy efficiency measures applied together in order to weatherproof a building. A weatherisation package might include, for example, the retrofitting of insulation and draught-proofing doors and windows.

Well-being refers to the integrated physiological, psychological and mental state of an individual, a household or group of people. It is broader than health, which typically refers to the physical state of an individual, family or group of people (public health).

© OECD/IEA, 2014.

Acronyms, abbreviations and units

Acronyms and abbreviations

BAU	business as usual
BCA	benefit-cost analysis
BEAR	Berkeley Energy and Resources (model)
CEA	cost-effectiveness analysis
CFC	chlorofluorocarbon
CGE	computable general equilibrium (model)
CNG	compressed natural gas
CO	carbon monoxide
CO_2	carbon dioxide
CPI	consumer price index
CPUC	California Public Utilities Commission
CSC	conservation supply curve
CSR	corporate sustainability reporting
DALY	disability-adjusted life year
DPU	Department of Public Utilities (Massachusetts)
DRIPE	demand-reduction induced price effect
DSM	demand-side management
E3ME	Energy-Environment-Economy Macro-econometric Model for Europe
EE	energy efficiency
EEO	energy efficiency obligation
EMAK	Energy Management Action Network
ENE	Environment Northeast
EnMS	energy management system
ERR	economic rate of return
ESA	Energy Savings Assistance (Californian programme)
ETS	emissions trading scheme
EU	European Union
EU-15	European Union prior to the accession of ten candidate countries on 1 May 2004
EU-27	27 member states of the European Union prior to the accession of Croatia on 1 July 2013
EU-ETS	European Union Emissions Trading Scheme

© OECD/IEA, 2014.

EWM	excess winter mortality
EWS	Efficient World Scenario
FBI	Federal Buildings Initiative
GDP	gross domestic product
GHG	greenhouse gas
GNP	gross national product
GSEP	Global Superior Energy Performance
GVA	gross value added
HFC	hydrofluorocarbon
HHSRS	Housing Health and Safety Rating System
HIA	health impact assessment
HIDEEM	Health Impacts of Domestic Energy Efficiency Measures (model)
ICT	information and communication technology
IEA	International Energy Agency
IRR	internal rate of return
I-O	input-output (model)
IT	information technology
KfW	Kreditanstalt für Wiederaufbau
KPI	key performance indicator
LCA	lifecycle cost assessment
LIHEAP	Low Income Home Energy Assistance Program
MCDA	multi-criteria decision analysis
NBM	Net Benefit Model
NEB	non-energy benefit
NO_x	nitrous oxides
NPV	net present value
O&M	operation and maintenance
OECD	Organisation for Economic Co-operation and Development
ORNL	Oak Ridge National Laboratory
PCGE	partial computable general equilibrium
PED	primary energy demand
QALY	quality-adjusted life years
REC	renewable energy certificate
RGGI	Regional Greenhouse Gas Initiative
RPS	Renewable Portfolio Standards
SCT	Societal Cost Test
SDG&E	San Diego Gas & Electric
SDM	system dynamics modelling
SE4ALL	Sustainable Energy for All Initiative (UN)
SEAI	Sustainable Energy Authority of Ireland
SEP	Superior Energy Performance

© OECD/IEA, 2014.

SF-36	Health Survey
SO_x	sulphur oxides
T&D	transmission and distribution
TRC	Total Resource Cost Test
UCL	University College London
UK	United Kingdom
UN	United Nations
UNFCCC	United Nations Framework Convention on Climate Change
US	United States
US EPA	United States Environmental Protection Agency
VAT	value-added tax
WAP	Weatherization Assistance Program (United States)
WUNZ: HS	Warm Up New Zealand: Heat Smart
WWF	World Wild Fund for Nature

Units

°C	degree Celsius
CAD	Canadian dollar
CNY	Chinese yuan renminbi
EUR	euro
EUR/Mtoe	euros per million tonnes of oil-equivalent
EUR/MWh	euros per megawatt hour
GBP	British pound
GBP/MWh	British pounds per megawatt hour
kt	kilotonne
kW	kilowatt
kWh/yr	kilowatt hours per year
Mtoe	million tonnes of oil-equivalent
MWh	megawatt hour
NZD	New Zealand dollar
PLN	Polish zloty
tCO_2	tonne of carbon dioxide
UAH	Ukrainian hryvnia
USD	United States dollar
USD/bbl	United States dollars per barrel (of oil)
USD/kWh/yr	United States dollars per kilowatt hour per year
USD/MWh	United States dollars per megawatt hour
USD/yr	United States dollars per year

© OECD/IEA, 2014.

List of boxes

Introduction 7

Box ES.1 Energy efficiency generates important benefits for emerging economies 23

Chapter 1 Taking a multiple benefits approach to energy efficiency 27

Box 1.1 Energy efficiency as a tool to reduce GHG emissions 30
Box 1.2 Energy efficiency as a tool for development in the UN SE4ALL initiative 32
Box 1.3 How the IEA selected focus areas in this publication 34
Box 1.4 Energy efficiency for emerging economies: "Doing even more with more" 35
Box 1.5 A key tool for tackling fuel poverty 36

Chapter 2 Macroeconomic impacts of energy efficiency 45

Box 2.1 Energy security 49
Box 2.2 Transformational effects 52
Box 2.3 Possible steps in an assessment of macroeconomic impacts of energy efficiency measures 58

Chapter 3 Public budget impacts of energy efficiency 71

Box 3.1 Obstacles to including public budget impacts in BCA 72
Box 3.2 Estimating how energy efficiency measures affect revenue from sales taxes 74
Box 3.3 Estimating the expenditure of a loan programme 76
Box 3.4 Estimates of employment impacts from energy efficiency measures 76
Box 3.5 Reducing public sector energy costs through energy efficiency 78
Box 3.6 Addressing the public costs of energy subsidies 80
Box 3.7 Changes in revenue from energy excise duty 81
Box 3.8 Net effects of changes to taxes on other goods and services 81
Box 3.9 Estimations of public health benefits 83
Box 3.10 Reducing future infrastructure costs through energy efficiency measures 84
Box 3.11 Estimation of public budget expenditure for an EU loan programme 88
Box 3.12 Input-output model to estimate public budget impacts in Germany 89

Chapter 4 Health and well-being impacts of energy efficiency 97

Box 4.1 IEA governments get serious about energy efficiency policies that target health and well-being 99
Box 4.2 Improvements in building insulation deliver greatest health impacts 101
Box 4.3 UK study produces concrete results on the mental health impacts of energy efficiency 106
Box 4.4 Air quality in the outdoor environment and health impacts 109
Box 4.5 Quantified indicators for impacts arising from energy efficiency improvements 112

© OECD/IEA, 2014.

Box 4.6 Using reduced absenteeism to calculate the financial value of improved health 113
Box 4.7 Modelling the health impacts of energy efficiency measures 115

Chapter 5 Industrial sector impacts of energy efficiency 129

Box 5.1 The primary objective: Industrial optimisation or upgrading projects? 131
Box 5.2 Key drivers for business-led energy efficiency 131
Box 5.3 Energy efficiency can reduce equipment damage 136
Box 5.4 Energy efficiency measures boost production in small manufacturing 137
Box 5.5 Early assessments of the multiple benefits of industrial energy efficiency 138
Box 5.6 Initiatives to promote data collection and sharing 139
Box 5.7 Consider confidentiality 140
Box 5.8 A company-level ex ante assessment of additional benefits 140
Box 5.9 An example of monetisation based on existing data 145
Box 5.10 A pragmatic step-by-step process for early investigations 145
Box 5.11 Australia and Sweden develop industrial energy efficiency programmes with multiple benefits in mind 147

Chapter 6 Energy delivery impacts of energy efficiency 153

Box 6.1 Good prospects for EEOs in both vertically integrated and liberalised markets 155
Box 6.2 Reducing energy sales still makes sense for energy providers 155
Box 6.3 Efficiency Vermont: An extensive assessment of energy provider multiple benefits 157
Box 6.4 Avoided system/network costs: An example from Con Edison in New York 160
Box 6.5 Germany: Quantifying system-level direct benefits 162
Box 6.6 Cost of saving energy and CO2 in the United Kingdom through EEOs 165
Box 6.7 Efforts to measure and evaluate NEBs in Massachusetts 168
Box 6.8 Demand-reduction induced price effects (DRIPES) 169
Box 6.9 Efficiency Vermont: Impact of end-user benefits on direct energy provider benefits 170
Box 6.10 A proposal for framework evolution: Resource Value Framework 172

Chapter 7 Conclusion: A new perspective on energy efficiency 179

Box 7.1 "Zip-Up" implementation method 182
Box 7.2 Getting the energy efficiency message across to households 184

Companion Guide Methodologies for the multiple benefits approach 189

Box 1 Trickle-through effects in the multiple benefits of energy efficiency 196
Box 2 A pioneering bottom-up model of net benefits 199
Box 3 Choosing the most appropriate assessment method 201
Box 4 Five cost-effectiveness tests expanding to include multiple benefits assessment 203
Box 5 Calculation methods in benefit-cost analysis 204
Box 6 A promising option: System dynamics modelling 205

© OECD/IEA, 2014.

List of figures

Introduction 7

Figure ES.1 Long-term energy efficiency economic potential by sector 19
Figure ES.2 The multiple benefits of energy efficiency improvements 20

Chapter 1 Taking a multiple benefits approach to energy efficiency 27

Figure 1.1 The multiple benefits of energy efficiency 28
Figure 1.2 Global levels of investment in selected areas of the energy system in 2011 31
Figure 1.3 Long-term energy efficiency economic potential by sector 33

Chapter 2 Macroeconomic impacts of energy efficiency 45

Figure 2.1 Energy demands reductions effects 47

Chapter 3 Public budget impacts of energy efficiency 71

Figure 3.1 Public budget impacts of energy efficiency measures 73

Chapter 4 Health and well-being impacts of energy efficiency 97

Figure 4.1 HIDEEM results for cost-effectiveness of energy efficiency for low-income households 115

Chapter 5 Industrial sector impacts of energy efficiency 129

Figure 5.1 Matrix classifying industrial benefits in terms of quantifiability and time horizon 142

Chapter 6 Energy delivery impacts of energy efficiency 153

Figure 6.1 Multiple benefits accrued to Vermont energy providers in 2010 158
Figure 6.2 Comparison of average and marginal line losses for typical US utility as a function of the load 161
Figure 6.3 Potential electricity systems savings in Germany: 2035 163
Figure 6.4 NEBs for utilities from low-income energy savings programmes 167

© OECD/IEA, 2014.

Companion Guide Methodologies for the multiple benefits approach 189

Figure 1	Methods to monetise outcomes	193
Figure 2	Simultaneous impacts of a building retrofit programme across sectors and economic levels	196
Figure 3	Methods to assess indirect impacts, accounting for scope and complexity	197
Figure 4	Decision tree	201
Figure 5	Schematic of interactions in a system comprising housing, energy and well-being	206

© OECD/IEA, 2014.

List of tables

Chapter 1 Taking a multiple benefits approach to energy efficiency 27

Table 1.1 The interaction between multiple benefits and energy savings 40

Chapter 2 Macroeconomic impacts of energy efficiency 45

Table 2.1 Overview of macroeconomic indicators for energy efficiency impact assessment 51
Table 2.2 Overview of several macroeconomic models of energy efficiency impacts 57
Table 2.3 Macroeconomic impacts included in energy efficiency policy appraisal 59
Table 2.4 A selection of estimates of macroeconomic impacts of energy efficiency programmes 62
Table 2.5 Further stakeholder research and collaboration opportunities in macroeconomic impacts 65

Chapter 3 Public budget impacts of energy efficiency 71

Table 3.1 Basic assessment of public budget impacts from energy efficiency programmes 86
Table 3.2 Fiscal multipliers for EU-27 87
Table 3.3 Declining long-term discount rate applied in the United Kingdom 91
Table 3.4 New Zealand discount rates 91
Table 3.5 Further stakeholder research and collaboration opportunities in public budget impacts 92

Chapter 4 Health and well-being impacts of energy efficiency 97

Table 4.1 Overview of direct and indirect impacts of improved energy efficiency on health and well-being 103
Table 4.2 Common indicators used in measuring health and well-being impacts of energy efficiency 112
Table 4.3 Further stakeholder research and collaboration opportunities in health and well-being 118

Chapter 5 Industrial sector impacts of energy efficiency 129

Table 5.1 Getting stakeholder buy-in to the business case for energy efficiency 133
Table 5.2 Company-level benefits from industrial energy efficiency projects 134
Table 5.3 Choosing a method, exact calculation or estimation, according to assessment needs 144
Table 5.4 Further research and collaboration opportunities in industrial sector impacts 148

Chapter 6 Energy delivery impacts of energy efficiency 153

Table 6.1 Energy provider multiple benefits arising from energy efficiency 156
Table 6.2 Multiple benefits accrued to Vermont energy providers in 2010 158
Table 6.3 Total annual costs for German power generation and grid infrastructure in 2035 162

© OECD/IEA, 2014.

Table 6.4 UK EEO assessment compares BAU against a proposed Central Scenario 165
Table 6.5 NEB values for participants of the SDG&E programme 168
Table 6.6 Multiple benefits accrued to Vermont energy providers from end-user benefits in 2010 170
Table 6.7 Further stakeholder research and collaboration opportunities in energy delivery impacts 174

Chapter 7 Conclusion: A new perspective on energy efficiency 179

Table 7.1 Illustrative valuation of benefits in relation to varying priorities of stakeholders 181
Table 7.2 Potential areas of interest among stakeholders in the multiple benefits approach 185

Companion Guide Methodologies for the multiple benefits approach 189

Table 1 Integrating the multiple benefits approach in an IEA Policy Pathway 207

© OECD/IEA, 2014.